TEACHER RESOURCES

Materials and Motion

FOSS

Next Generation

Full Option Science System
Developed at the Lawrence Hall of Science, University of California, Berkeley
Published and Distributed by Delta Education

FOSS Lawrence Hall of Science Team

Larry Malone and Linda De Lucchi, FOSS Project Codirectors and Lead Developers
Kathy Long, FOSS Assessment Director; David Lippman, Program Manager; Carol Sevilla, Publications Design Coordinator; Susan Stanley, Illustrator; John Quick, Photographer
FOSS Curriculum Developers: Brian Campbell, Teri Lawson, Alan Gould, Susan Kaschner Jagoda, Ann Moriarty, Jessica Penchos, Kimi Hosoume, Virginia Reid, Joanna Snyder, Erica Beck Spencer, Joanna Totino, Diana Velez, Natalie Yakushiji
Susan Ketchner, Technology Project Manager
FOSS Technology Team: Dan Bluestein, Christopher Cianciarulo, Matthew Jacoby, Kate Jordan, Frank Kusiak, Nicole Medina, Jonathan Segal, Dave Stapley, Shan Tsai

Delta Education Team

Bonnie A. Piotrowski, Editorial Director, Elementary Science
Project Team: Mathew Bacon, Jennifer Apt, Sandra Burke, Tom Guetling, Joann Hoy, Jacquelyn Lachance, Angela Miccinelloy

Thank you to all FOSS Grades K-5 Trial Teachers

Heather Ballard, Wilson Elementary, Coppell, TX; Mirith Ballestas De Barroso, Treasure Forest Elementary, Houston, TX; Terra L. Barton, Harry McKillop Elementary, Melissa, TX; Rhonda Bernard, Frances E. Norton Elementary, Allen, TX; Theresa Bissonnette, East Millbrook Magnet Middle School, Raleigh, NC; Peter Blackstone, Hall Elementary School, Portland, ME; Tiffani Brisco, Seven Hills Elementary, Newark, TX; Darrow Brown, Lake Myra Elementary School, Wendell, NC; Heather Callaghan, Olive Chapel Elementary, Apex, NC; Katie Cannon, Las Colinas Elementary, Irving, TX; Elaine M. Cansler, Brassfi eld Road Elementary School, Raleigh, NC; Kristy Cash, Wilson Elementary, Coppell, TX; Monica Coles, Swift Creek Elementary School, Raleigh, NC; Shirley Conner, Ocean Avenue Elementary School, Portland, ME; Sally Connolly, Cape Elizabeth Middle School, Cape Elizabeth, ME; Melissa Cook-Airhart, Harry McKillop Elementary, Melissa, TX; Melissa Costa, Olive Chapel Elementary, Apex, NC; Hillary P. Croissant, Harry McKillop Elementary, Melissa, TX; Rene Custeau, Hall Elementary School, Portland, ME; Nancy Davis, Martha and Josh Morriss Mathematics and Engineering Elementary School, Texarkana, TX; Nancy Deveneau, Wilson Elementary, Coppell, TX; Karen Diaz, Las Colinas Elementary, Irving, TX; Marlana Dumas, Las Colinas Elementary, Irving, TX; Mary Evans, R.E. Good Elementary School, Carrollton, TX; Jacquelyn Farley, Moss Haven Elementary, Dallas, TX; Corinna Ferrier, Oak Forest Elementary, Humble, TX; Allison Fike, Wilson Elementary, Coppell, TX; Barbara Fugitt, Martha and Josh Morriss Mathematics and Engineering Elementary School, Texarkana, TX; Colleen Garvey, Farmington Woods Elementary, Cary, NC; Judy Geller, Bentley Elementary School, Oakland, CA; Erin Gibson, Las Colinas Elementary, Irving, TX; Kelli Gobel, Melissa Ridge Intermediate School, Melissa, TX; Dollie Green, Melissa Ridge Intermediate School, Melissa, TX; Brenda Lee Harrigan, Bentley Elementary School, Oakland, CA; Cori Harris, Samuel Beck Elementary, Trophy Club, TX; Kim Hayes, Martha and Josh Morriss Mathematics and Engineering Elementary School, Texarkana, TX; Staci Lynn Hester, Lacy Elementary School, Raleigh, NC; Amanda Hill, Las Colinas Elementary, Irving, TX; Margaret Hillman, Ocean Avenue Elementary School, Portland, ME; Cindy Holder, Oak Forest Elementary, Humble, TX; Sarah Huber, Hodge Road Elementary, Knightdale, NC; Susan Jacobs, Granger Elementary, Keller, TX; Carol Kellum, Wallace Elementary, Dallas, TX; Jennifer A. Kelly, Hall Elementary School, Portland, ME; Brittani Kern, Fox Road Elementary, Raleigh, NC; Jodi Lay, Lufkin Road Middle School, Apex, NC; Melissa Lourenco, Lake Myra Elementary School, Wendell, NC; Ana Martinez, RISD Academy, Dallas, TX; Shaheen Mavani, Las Colinas Elementary, Irving, TX; Mary Linley McClendon, Math Science Technology Magnet School, Richardson, TX; Adam McKay, Davis Drive Elementary, Cary, NC; Leslie Meadows, Lake Myra Elementary School, Wendell, NC; Anne Mechler, J. Erik Jonsson Community School, Dallas, TX; Anne Miller, J. Erik Jonsson Community School, Dallas, TX; Shirley Diann Miller, The Rice School, Houston, TX; Keri Minier, Las Colinas Elementary, Irving, TX; Stephanie Renee Nance, T.H. Rogers Elementary, Houston, TX; Cynthia Nilsen, Peaks Island School, Peaks Island, ME; Elizabeth Noble, Las Colinas Elementary, Irving, TX; Courtney Noonan, Shadow Oaks Elementary School, Houston, TX; Sarah Peden, Aversboro Elementary School, Garner, NC; Carrie Prince, School at St. George Place, Houston, TX; Marlaina Pritchard, Melissa Ridge Intermediate School, Melissa, TX; Alice Pujol, J. Erik Jonsson Community School, Dallas, TX; Claire Ramsbotham, Cape Elizabeth Middle School, Cape Elizabeth, ME; Paul Rendon, Bentley Elementary, Oakland, CA; Janette Ridley, W.H. Wilson Elementary School, Coppell, TX; Kristina (Crickett) Roberts, W.H. Wilson Elementary School, Coppell, TX; Heather Rogers, Wendell Creative Arts & Science Magnet Elementary School, Wendell, NC; Alissa Royal, Melissa Ridge Intermediate School, Melissa, TX; Megan Runion, Olive Chapel Elementary, Apex, NC; Christy Scheef, J. Erik Jonsson Community School, Dallas, TX; Samrawit Shawl, T.H. Rogers School, Houston, TX; Nicole Spivey, Lake Myra Elementary School, Wendell, NC; Ashley Stephenson, J. Erik Jonsson Community School, Dallas, TX; Jolanta Stern, Browning Elementary School, Houston, TX; Gale Stimson, Bentley Elementary, Oakland, CA; Ted Stoeckley, Hall Middle School, Larkspur, CA; Cathryn Sutton, Wilson Elementary, Coppell, TX; Camille Swander, Ocean Avenue Elementary School, Portland, ME; Brandi Swann, Westlawn Elementary School, Texarkana, TX; Robin Taylor, Arapaho Classical Magnet, Richardson, TX; Michael C. Thomas, Forest Lane Academy, Dallas, TX; Jomarga Thompkins, Lockhart Elementary, Houston, TX; Mary Timar, Madera Elementary, Lake Forest, CA; Helena Tongkeamha, White Rock Elementary, Dallas, TX; Linda Trampe, J. Erik Jonsson Community School, Dallas, TX; Charity VanHorn, Fred A. Olds Elementary, Raleigh, NC; Kathleen VanKeuren, Lufkin Road Middle School, Apex, NC; Valerie Vassar, Hall Elementary School, Portland, ME; Megan Veron, Westwood Elementary School, Houston, TX; Mary Margaret Waters, Frances E. Norton Elementary, Allen, TX; Stephanie Robledo Watson, Ridgecrest Elementary School, Houston, TX; Lisa Webb, Madisonville Intermediate, Madisonville, TX; Matt Whaley, Cape Elizabeth Middle School, Cape Elizabeth, ME; Nancy White, Canyon Creek Elementary, Austin, TX; Barbara Yurick, Oak Forest Elementary, Humble, TX; Linda Zittel, Mira Vista Elementary, Richmond, CA

Photo Credits: © OlegD/Shutterstock (cover); © Zach Smith; © Monkey Business Images/Shutterstock; © David Lippman; © Christian Musat/Shutterstock

Published and Distributed by Delta Education, a member of the School Specialty Family
The FOSS program was developed in part with the support of the National Science Foundation grant nos. MDR-8751727 and MDR-9150097. However, any opinions, findings, conclusions, statements, and recommendations expressed herein are those of the authors and do not necessarily reflect the views of NSF. FOSSmap was developed in collaboration between the BEAR Center at UC Berkeley and FOSS at the Lawrence Hall of Science.

Standards cited herein from NGSS Lead States. 2013. *Next Generation Science Standards: For States, By States.* Washington, DC: The National Academies Press. Next Generation Science Standards is a registered trademark of Achieve. Neither Achieve nor the lead states and partners that developed the Next Generation Science Standards was involved in the production of, and does not endorse, this product.

Materials and Motion — Teacher Toolkit, 1487670
Teacher Resources, 1487604
978-1-62571-425-1
Printing 5 – 4/2018
Patterson Printing, Benton Harbor, MI

TEACHER RESOURCES

TABLE OF CONTENTS

This document, *Teacher Resources*, is one of three parts of the *FOSS Teacher Toolkit* for this module. The chapters in *Teacher Resources* are all available as PDFs on FOSSweb.

The other parts of the module *Teacher Toolkit* are the *Investigations Guide* and a copy of the *FOSS Science Resources* student book containing original readings for this module.

The spiral-bound *Investigations Guide* contains these chapters.

- Overview
- Framework and NGSS
- Materials
- Technology
- Investigations
- Assessment

The *Teacher Toolkit* is the most important part of the FOSS Program. It is here that all the wisdom and experience contributed by hundreds of educators has been assembled. Everything we know about the content of the module, how to teach the subject, and the resources that will assist the effort are presented here.

FOSS Program Goals

FOSS Program Goals

Contents

INTRODUCTION

The Full Option Science System™ has evolved from a philosophy of teaching and learning at the Lawrence Hall of Science that has guided the development of successful active-learning science curricula for more than 40 years. The FOSS Program bridges research and practice by providing tools and strategies to engage students and teachers in enduring experiences that lead to deeper understanding of the natural and designed worlds.

Science is a creative and analytic enterprise, made active by our human capacity to think. Scientific knowledge advances when scientists observe objects and events, think about how they relate to what is known, test their ideas in logical ways, and generate explanations that integrate the new information into understanding of the natural world. Engineers apply that understanding to solve real-world problems. Thus, the scientific enterprise is both what we know (content knowledge) and how we come to know it (practices). Science is a discovery activity, a process for producing new knowledge.

The best way for students to appreciate the scientific enterprise, learn important scientific and engineering concepts, and develop the ability to think well is to actively participate in scientific practices through their own investigations and analyses. FOSS was created to engage students and teachers with meaningful experiences in the natural and designed worlds.

GOALS OF THE FOSS PROGRAM

FOSS has set out to achieve three important goals: scientific literacy, instructional efficiency, and systemic reform.

Scientific Literacy

FOSS provides all students with science experiences that are appropriate to students' cognitive development and prior experiences. It provides a foundation for more advanced understanding of core science ideas which are organized in thoughtfully designed learning progressions and prepares students for life in an increasingly complex scientific and technological world.

The National Research Council (NRC) in *A Framework for K–12 Science Education* and the American Association for the Advancement of Science (AAAS) in *Benchmarks for Scientific Literacy,* have described the characteristics of scientific literacy:

- Familiarity with the natural world, its diversity, and its interdependence.
- Understanding the disciplinary core ideas and the cross-cutting concepts of science, such as patterns; cause and effect; scale, proportion, and quantity; systems and system models; energy and matter—flows, cycles, and conservation; structure and function; and stability and change.
- Knowing that science and engineering, technology, and mathematics are interdependent human enterprises and, as such, have implied strengths and limitations.
- Ability to reason scientifically.
- Using scientific knowledge and scientific and engineering practices for personal and social purposes.

The FOSS Program design is based on learning progressions that provide students with opportunities to investigate core ideas in science in increasingly complex ways over time. FOSS starts with the intuitive ideas that primary students bring with them and provides experiences that allow students to develop more sophisticated understanding as they grow through the grades. Cognitive research tells us that learning involves individuals in actively constructing schemata to organize new information and to relate and incorporate the new understanding into established knowledge. What sets experts apart from novices is that experts in a discipline have extensive knowledge that is effectively organized into structured schemata to promote thinking. Novices have disconnected ideas about a topic that are difficult to retrieve and use. Through internal processes to establish schemata and through social processes of interacting with peers and adults, students construct understanding of the natural world and their relationship to it. The target goal for FOSS students is to know and use scientific explanations of the natural world and the designed world; to understand the nature and development of scientific knowledge and technological capabilities; and to participate productively in scientific and engineering practices.

Instructional Efficiency

FOSS provides all teachers with a complete, cohesive, flexible, easy-to-use science program that reflects current research on teaching and learning, including student discourse, argumentation, writing to learn, and reflective thinking, as well as teacher use of formative assessment to guide instruction. The FOSS Program uses effective instructional methodologies, including active learning, scientific practices, focus questions to guide inquiry, working in collaborative groups, multisensory strategies, integration of literacy, appropriate use of digital technologies, and making connections to students' lives, including the outdoors.

FOSS is designed to make active learning in science engaging for teachers as well as for students. It includes these supports for teachers:

- Complete equipment kits with durable, well-designed materials for all students.
- Detailed *Investigations Guide* with science background for the teacher and focus questions to guide instructional practice and student thinking.
- Multiple strategies for formative assessment at all grade levels.
- Benchmark assessments (grades 1–5) with online access for administering, coding, and analyzing assessments (grades 3–5).
- Strategies for use of science notebooks for novice and experienced users.
- *FOSS Science Resources,* a book of module-specific readings with strategies for science-centered language development.
- The FOSS website with interactive multimedia activities for use in school or at home, suggested interdisciplinary-extension activities, and extensive online support for teachers, including teacher prep videos.

FOSS Program Goals

Systemic Reform

FOSS provides schools and school systems with a program that addresses the community science-achievement standards. The FOSS Program prepares students by helping them acquire the knowledge and thinking capacity appropriate for world citizens.

The FOSS Program design makes it appropriate for reform efforts on all scales. It reflects the core ideas to be incorporated into the next-generation science standards. It meets with the approval of science and technology companies working in collaboration with school systems, and it has demonstrated its effectiveness with diverse student and teacher populations in major urban reform efforts. The use of science notebooks and formative-assessment strategies in FOSS redefines the role of science in a school—the way that teachers engage in science teaching with one another as professionals and with students as learners, and the way that students engage in science learning with the teacher and with one another. FOSS takes students and teachers beyond the classroom walls to establish larger communities of learners.

BRIDGING RESEARCH INTO PRACTICE

The FOSS Program is built on the assumptions that understanding core scientific knowledge and how science functions is essential for citizenship, that all teachers can teach science, and all that students can learn science. The guiding principles of the FOSS design, described below, are derived from research and confirmed through FOSS developers' extensive experience with teachers and students in typical American classrooms.

Understanding of science develops over time. FOSS has elaborated learning progressions for core ideas in science for kindergarten through grade 6. Developing the learning progressions involves identifying successively more sophisticated ways of thinking about core ideas over multiple years. "If mastery of a core idea in a science discipline is the ultimate educational destination, then well-designed learning progressions provide a map of the routes that can be taken to reach that destination" (National Research Council, *A Framework for K–12 Science Education*, 2012).

Focusing on a limited number of topics in science avoids shallow coverage and provides more time to explore core science ideas in depth. Research emphasizes that fewer topics experienced in greater depth produces much better learning than many topics briefly visited. FOSS affirms this research. FOSS modules provide long-term engagement (9–10 weeks) with important science ideas. Furthermore, modules build upon one another within and across each strand, progressively moving students toward the grand ideas of science. The core ideas of science are difficult and complex, never learned in one lesson or in one class year.

FOSS Next Generation—Elementary Module Sequences

	PHYSICAL SCIENCE		EARTH SCIENCE		LIFE SCIENCE	
	MATTER	ENERGY AND CHANGE	DYNAMIC ATMOSPHERE	ROCKS AND LANDFORMS	STRUCTURE/ FUNCTION	COMPLEX SYSTEMS
5	Mixtures and Solutions		Earth and Sun		Living Systems	
4		Energy		Soils, Rocks, and Landforms	Environments	
3	Motion and Matter		Water and Climate		Structures of Life	
2	Solids and Liquids			Pebbles, Sand, and Silt	Insects and Plants	
1		Sound and Light	Air and Weather		Plants and Animals	
K	Materials and Motion		Trees and Weather		Animals Two by Two	

FOSS Program Goals

Science is more than a body of knowledge. How well you think is often more important than how much you know. In addition to the science content framework, every FOSS module provides opportunities for students to engage in and understand science practices, and many modules explore issues related to engineering practices and the use of natural resources. FOSS promotes these science and engineering practices described in *A Framework for K–12 Science Education.*

- Asking questions (for science) and defining problems (for engineering)
- Developing and using models
- Planning and carrying out investigations
- Analyzing and interpreting data
- Using mathematics and computational thinking
- Constructing explanations (for science) and designing solutions (for engineering)
- Engaging in argument from evidence
- Obtaining, evaluating, and communicating information

Science is inherently interesting, and children are natural investigators. It is widely accepted that children learn science concepts best by doing science. Doing science means hands-on experiences with objects, organisms, and systems. Hands-on activities are motivating for students, and they stimulate inquiry and curiosity. For these reasons, FOSS is committed to providing the best possible materials and the most effective procedures for deeply engaging students with scientific concepts. FOSS students at all grade levels investigate, experiment, gather data, organize results, and draw conclusions based on their own actions. The information gathered in such activities enhances the development of scientific and engineering practices.

Education is an adventure in self-discovery. Science provides the opportunity to connect to students' interests and experiences. Prior experiences and individual learning styles are important considerations for developing understanding. Observing is often equated with seeing, but in the FOSS Program all senses are used to promote greater understanding. FOSS evolved from pioneering work done in the 1970s with students with disabilities. The legacy of that work is that FOSS investigations naturally use multisensory methods to accommodate students with physical and learning disabilities and also to maximize information gathering for all students. A number of tools, such as the FOSS syringe and balance, were originally designed to serve the needs of students with disabilities.

Formative assessment is a powerful tool to promote learning and can change the culture of the learning environment. Formative assessment in FOSS creates a community of reflective practice. Teachers and students make up the community and establish norms of mutual support, trust, respect, and collaboration. The goal of the community is that everyone will demonstrate progress and will learn and grow.

Science-centered language development promotes learning in all areas. Effective use of science notebooks can promote reflective thinking and contribute to life long learning. Research has shown that when language-arts experiences are embedded within the context of learning science, students improve in their ability to use their language skills. Students are eager to read to find out information, and to share their experiences both verbally and in writing.

Experiences out of the classroom develop awareness of community. By extending classroom learning into the outdoors, FOSS brings the science concepts and principles to life. In the process of validating classroom learning among the schoolyard trees and shrubs, down in the weeds on the asphalt, and in the sky overhead, students will develop a relationship with nature. It is our relationship with natural systems that allows us to care deeply for these systems.

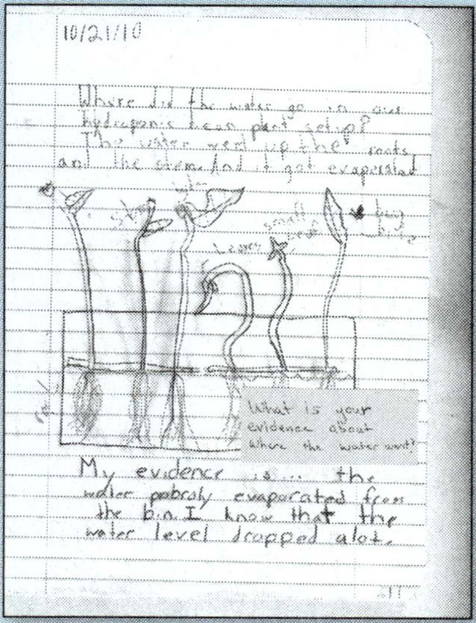

FOSS NEXT GENERATION K–8 SCOPE AND SEQUENCE

Grade	Physical Science	Earth Science	Life Science
6–8	Waves* Gravity and Kinetic Energy*	Planetary Science	Heredity and Adaptation* Human Systems Interactions*
	Chemical Interactions	Earth History	Populations and Ecosystems
	Electromagnetic Force* Variables and Design*	Weather and Water	Diversity of Life
5	Mixtures and Solutions	Earth and Sun	Living Systems
4	Energy	Soils, Rocks, and Landforms	Environments
3	Motion and Matter	Water and Climate	Structures of Life
2	Solids and Liquids	Pebbles, Sand, and Silt	Insects and Plants
1	Sound and Light	Air and Weather	Plants and Animals
K	Materials and Motion	Trees and Weather	Animals Two by Two

*** Half-length course**

FOSS is a research-based science curriculum for grades K–8 developed at the Lawrence Hall of Science, University of California, Berkeley. FOSS is also an ongoing research project dedicated to improving the learning and teaching of science. The FOSS project began over 25 years ago during a time of growing concern that our nation was not providing young students with an adequate science education. The FOSS Program materials are designed to meet the challenge of providing meaningful science education for all students in diverse American classrooms and to prepare them for life in the 21st century. Development of the FOSS Program was, and continues to be, guided by advances in the understanding of how people think and learn.

With the initial support of the National Science Foundation and continued support from the University of California, Berkeley, and School Specialty, Inc., the FOSS Program has evolved into a curriculum for all students and their teachers, grades K–8. The current editions of FOSS are the result of a rich collaboration among the FOSS/Lawrence Hall of Science development staff; the FOSS product development team at School Specialty; assessment specialists, educational researchers, and scientists; and dedicated professionals in the classroom and their students, administrators, and families.

We acknowledge the thousands of FOSS educators who have embraced the notion that science is an active process, and we thank them for their significant contributions to the development and implementation of the FOSS Program.

Science Notebooks in Grades K-2

Science Notebooks in Grades K–2

Student notebook from the Trees and Weather Module

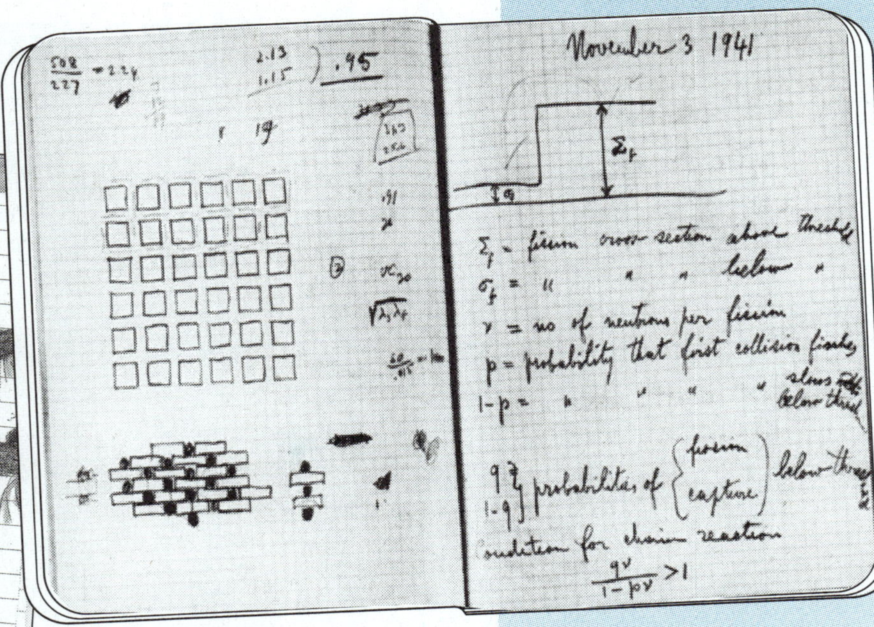

Enrico Fermi's laboratory notebook, 1941

Source: Special Collections Research Center, University of Chicago Library

INTRODUCTION

A scientist's notebook is a detailed record of his or her engagement with scientific phenomena. It is a personal representation of experiences, observations, and thinking—an integral part of the process of doing scientific work. A scientist's notebook is a continuously updated history of the development of scientific knowledge and reasoning. FOSS students are young scientists; they incorporate notebooks into their science learning.

This chapter is designed to be a resource for teachers who are incorporating notebooks into their classroom practice. For teachers just beginning to use notebooks, the Getting Started section in this chapter suggests how to set up the notebooks, and the *Investigations Guide* cues you when to engage students with the notebooks during the investigation. For more information on specific types of notebook entries, the subsections in the Notebook Components sections include strategies to differentiate instruction for various ability levels.

Contents

NOTEBOOK BENEFITS

Engaging in active science is one part experience and two parts making sense of the experience. Science notebooks help students with the sense-making part. Science notebooks assist with documentation and cognitive engagement. For teachers, notebooks are tools for gaining insight into students' thinking. Notebooks inform and refine instructional practice.

Benefits to Students
- *Documentation*
- *Reference document*
- *Cognitive engagement*

Benefits to Students

Documentation. Science provides an authentic experience for students to develop their documentation skills. In the primary classroom, getting students to use a notebook will introduce the powerful skills of information organization. Students document their experiences, data, and thinking during each investigation. They create simple tables, graphs, charts, drawings, and labeled illustrations as standard means for representing and displaying data. At first, students will look at their science notebooks as little more than a random collection of words and pictures. Each notebook page represents an isolated activity. As students become more accomplished at keeping notebooks, their documentation will become better organized and efficient. In time and with some guidance, students will adopt a deeper understanding of their collections as integrated records of their learning.

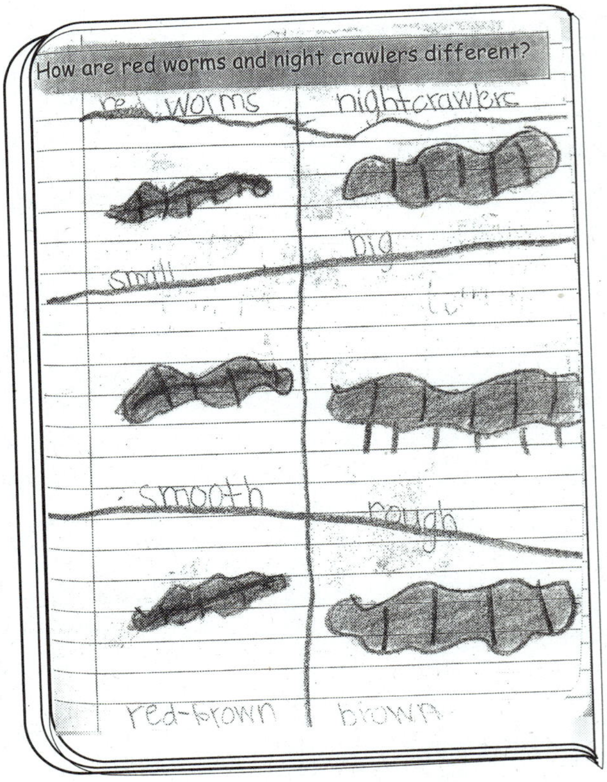

Kindergarten entry from the Animals Two by Two Module

Reference document. When data are displayed in functional ways, students can think about the data more effectively. Even with young students, a well-kept notebook is a useful reference document. When students have forgotten a fact about an insect or a plant that they learned earlier in their studies, they can look it up. Learning to trust a personal record of previous discoveries and knowledge structures is important.

A complete and accurate record allows students to reconstruct the sequence of learning events to "relive" the experience. Discussions about science among students; students and teachers; or students, teachers, and families have more meaning when they are supported by authentic documentation in students' notebooks.

Labeled drawing from the Animals Two by Two Module

Cognitive engagement. Once data are recorded and organized in an efficient manner in science notebooks, students can think about the data to draw conclusions about the way the world works. Their data, based on their experiences and observations, are the raw materials that students use to forge concepts and relationships.

Responding to a focus question from the Animals Two by Two Module with a drawing and words

Benefits to Teachers

In FOSS, the unit of instruction is the module—a sequence of conceptually related learning experiences that leads to a set of learning outcomes. A science notebook helps you think about and communicate the conceptual structure of the module you are teaching.

Assessment. From the assessment point of view, a science notebook is a collection of student-generated artifacts that exhibit learning. You can assess student skills, such as using drawings to record data, while students are working with materials. At other times, you collect the notebooks and review them in greater detail. The displays of data and analytical work, such as responses to focus questions, provide a measure of the quality and quantity of student learning. The notebook itself should not be graded. However, the notebook can be considered as one component of a student's overall performance in science.

Medium for feedback. The science notebook is an excellent medium for providing feedback to individual students regarding their work. Some students may be able to read a teacher comment written on a self-stick note, think about the issue, and respond. Other students may need oral feedback individually or in a small-group situation. This feedback might include additional modeling to help students make more accurate drawings, revisiting some important vocabulary, or introducing strategies to help students better communicate their thinking.

Focus for professional discussions. The science notebook acts as a focal point for discussion about students' learning at several levels. It can be reviewed and discussed during parent conferences. Science notebooks can be the focus of three-way discussions among students, teachers, and principals to ensure that all members of the school science community agree about what kinds of student work are valued and what level of performance to expect. Science notebooks shared among teachers in a study group or other professional-development environment can serve as a reflective tool that informs teachers of students' ability to demonstrate recording techniques, individual styles, various levels of good-quality work, and so on. Just as students can learn notebook strategies from one another, teachers can learn notebook skills from one another.

Benefits to Teachers

- *Assessment*
- *Medium for feedback*
- *Focus for professional discussions*
- *Refinement of practice*

A teacher provides feedback to a student.

Refinement of practice. As teachers, we are constantly looking for ways to improve instructional practices to increase students' understanding. Your use of the notebook should change over time. In the beginning, the focus will be on the notebook itself—what it looks like, what goes in it. As you become more comfortable with the notebook, the attention shifts to what students are learning. When this happens, you begin to consider how much scaffolding to provide to different students, how to use evidence of learning to differentiate instruction, and how to modify instruction to refine students' understanding.

Kindergarten entry from the Animals Two by Two Module

GETTING STARTED

Starting in kindergarten, students are expected to make detailed, thoughtful records of their science inquiries. While this may seem like a lofty goal, with some patience and thoughtful support, both teachers and students can learn how to use science notebooks effectively.

A major goal for using notebooks is to establish habits that will enable students to collect data and make sense of them. Use of the notebook must be flexible enough to allow students room to grow and supportive enough for students to be successful from the start. The format should be simple and the information meaningful to students. The notebook includes student drawings, simple writing in the form of individual words and short phrases, and a variety of visual and tactile artifacts. When students thumb through their notebooks, they are reminded of the objects and organisms they observed and their interactions with them.

Notebook Format

The *Teacher Resources* component of the FOSS *Teacher Toolkit* includes duplication masters for the same notebook sheets that are bound into the consumable notebooks. You can use these sheets to prepare an analogous notebook or to design a customized version.

In an autonomous approach, students create their entire science notebooks from blank pages in bound composition books. Students can glue or tape the provided notebook sheets into their notebooks, as well as create their own notebook entries. This level of notebook use will not be realized quickly and will require modeling by the teacher to provide enough structure to make the notebook useful. It will likely require systematic development by an entire teaching staff over several years.

You might choose to have a separate notebook for each module or one notebook for the entire year. (See the sidebar for the advantages of each.) Students will need about 30 pages (60 sides) for a typical module.

Advantages of One Notebook per Module:

- *Easy to replace if lost*
- *Lower cost*
- *Fewer pages*

Advantages of One Notebook per Year:

- *Easy to refer to prior activities*
- *Easy to see growth over time*

Organization of Notebooks

Two organizational components of the notebook should be planned right from the outset for primary students—page numbering and documentation. Two additional organizational components—a table of contents and an index—might be included only in the class notebook (detailed in the Supporting Students section) or in student notebooks toward the end of second grade. Not all organizational structures need to be present in individual student notebooks in the primary grades.

Page numbering. Each page should have a number. These can be applied to the pages, front to back, and referenced in the table of contents as the notebook progresses, or small blocks of pages can be prenumbered (pages 1–5 initially, pages 6–10 later, and so on) at appropriate times in the module.

A parent volunteer could number the pages. If students number them, monitor their work to make sure they don't skip pages or misnumber them.

Date, title, and other conditions. Each time students make a new entry, they should record certain information. At the very minimum, they should record the date and a title. More complete documentation might include the time; day of the week; team members; and if appropriate, weather conditions. When introducing a new condition to students, such as recording the names of team members, it is important to discuss why students are recording the information so that they understand the relevancy.

Some classes start each new entry at the top of the next available page. Others simply leave a modest space and enter the information right before the new entry.

Table of contents. If students keep a table of contents in their own notebooks, they should reserve the first two pages of the notebook for it. You will need to remind students to add to it systematically as you proceed through the module. The table of contents can be based on the names of the investigations in the module, the specific activities undertaken, the concepts learned, or some other schema that makes sense to everyone.

You might provide students with a preprinted table of contents without page numbers. As students work through each investigation, they record the relevant page numbers in their table of contents. If students are making their own table of contents, they should keep it fairly simple.

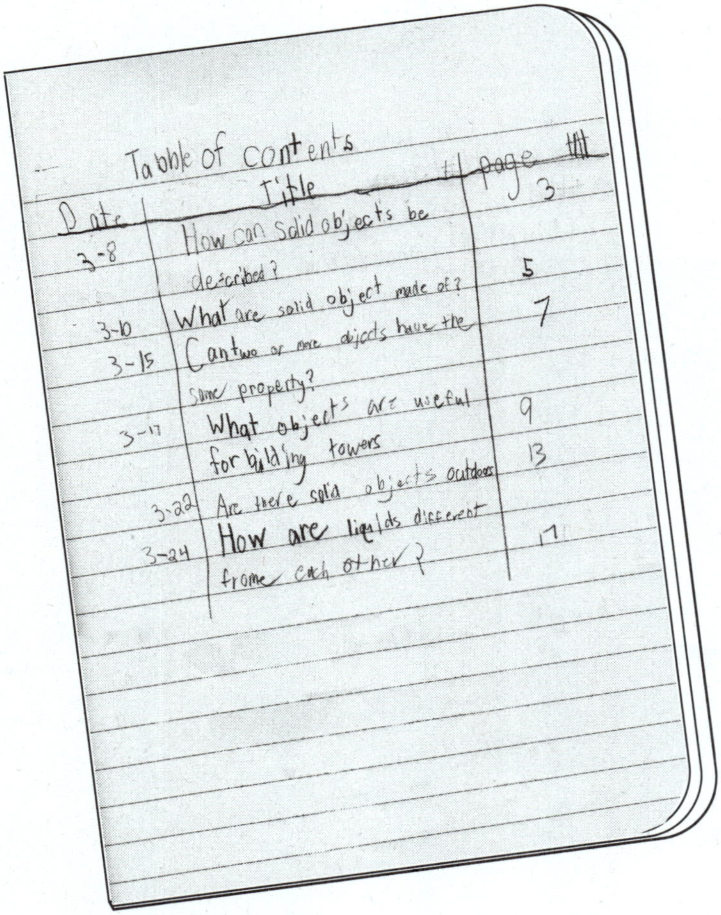

A table of contents for the Solids and Liquids Module

Index. Scientific academic language is important. FOSS strives to have students use precise, accurate vocabulary at all times in their writing and conversations. Key words can be displayed using a pocket chart, word wall, or written on index cards kept in boxes at each table. Another support to assist with acquisition of the scientific vocabulary is to introduce an index in the class notebook. It is not usually possible to enter the words in alphabetical order, since they will be acquired as the module advances. Instead, assign a block of letters to each of several index pages at the back of the class notebook (A–E, F–K, etc.); you or students can enter keys words. Students write the new vocabulary word or phrase in the appropriate square and tag it with the number of the page on which the word is defined in the notebook.

Including an index in the class notebook is a long-term time investment and can be overwhelming for the beginning notebook user. It may be better to forgo the index and use just a word wall to work with vocabulary initially.

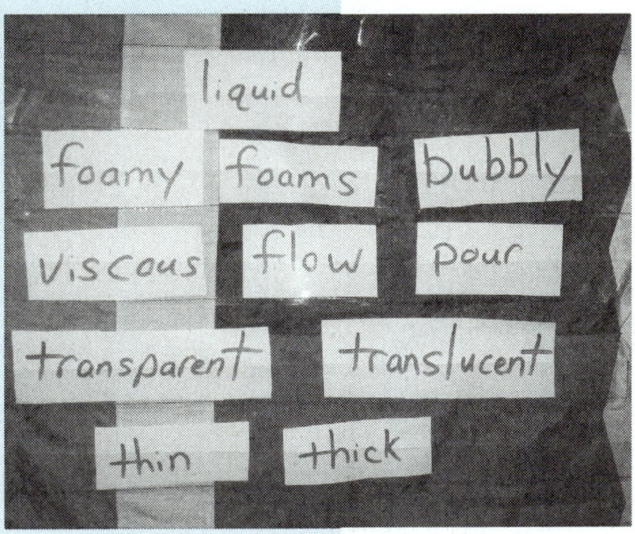

Pocket charts can be used to introduce vocabulary for students to include in their notebooks.

Notebook Entries

As students engage in scientific exploration, they will make entries in their notebooks. They might use a prepared notebook sheet or a more free-form entry. Students frequently respond to a focus question with a drawing or a simple written entry. Kindergartners may write single words; first and second graders will write simple observations and summary ideas, using the new vocabulary in their entries. These notebook entries allow early-childhood students to relive and describe their science experiences as they turn the pages in their notebooks.

Typically, the rules of grammar and spelling are relaxed when making notebook entries so as not to inhibit the flow of creative expression. Encourage students to use many means of recording and communicating besides writing, including charts, graphs, drawings, color codes, numbers, and images attached to the notebook pages. By exploring many options for making notebook entries, each student will find his or her most efficient, expressive way to capture and organize information for later retrieval.

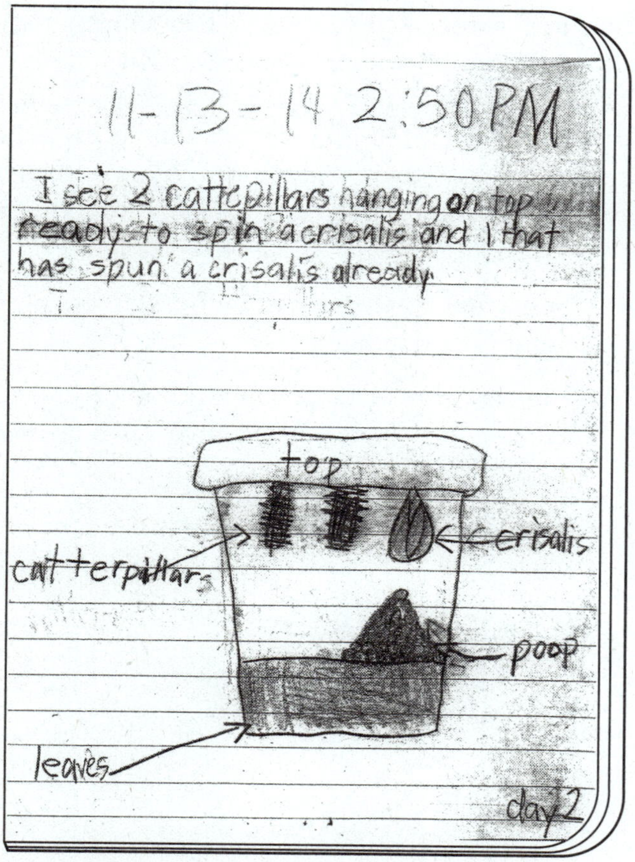

A notebook entry from the Insects and Plants Module

Supporting Students

Elementary classrooms contain students with a range of abilities, which is important to take into account when thinking about strategies for implementing science notebooks. Students need to have successful early experiences with notebooks. This requires planning. For younger students, a blank notebook may be intimidating, and they will look to you for guidance. FOSS teachers have had success using different supports and scaffolds to help transform the blank notebook into a valuable reference tool.

TEACHING NOTE

Use of the class notebook should be thoughtfully timed. Doing a class-notebook entry at the end of an activity is helpful to teach the components of a notebook, yet allows you to see what students do on their own. If you want to model a specific notebook strategy, use the class notebook during the activity.

Class notebook. You can create a class notebook to document scientific explorations as a way to model the various notebook components. Use a chart-paper tablet so that the pages can be flipped back and forth, and make it accessible at all times as a student reference. Students can emulate the class notebook in their own personal notebooks. Initially, individual notebooks will look quite similar to the class notebook. It is not the intent that students' notebooks be identical to the class notebook.

Scaffolds. Supports and scaffolds differ in one way. Supports are always available for students to access, such as allowing students access to a class notebook. Scaffolds are available just when the student needs them and will vary from student to student and from investigation to investigation. Scaffolds are meant to provide structure to a notebook entry and allow students to insert their observations into that structure. As the year progresses, the scaffolds change to allow for more student initiative. Scaffolds include

- **Sentence starters** or **drawing starters** provide a beginning point for a notebook entry.
- **Frames** provide more support but leave specific gaps for students to complete. Here's an example: "We planted _____ seeds and used _____ to water them."
- The suggested **notebook sheets** can guide students to record observations and data. The notebook sheets also guide thinking.

Think-alouds. Think-alouds help explain the decision-making process practiced by a savvy notebook user. They verbalize the thoughts used to create a particular notebook entry. For example, if students have recorded observations about one type of rock and are going to observe a second type, you might say,

I am going to observe another kind of rock. I'm going to look back to see how I recorded rock information before. I see that I made a large, detailed drawing. I described the texture, size, and color of the rock, too. So now I think I will make a drawing of the new rock. I'm going to record the texture, size, and color, too.

Now I know that one way I can get ideas for what to write in my notebook is to look back at observations I wrote before.

Providing time to record. When young students are engaged in active science, their efforts are focused on the materials, not the notebook. Students need this time to explore, and many will not open their notebooks and record observations, even with prompting. Students need separate time to record observations that fully document their discoveries. Some teachers have found it easier to leave the materials on the table and have students bring their notebooks to a common writing area. Then the teacher revisits the focus question or task and provides a few minutes for students to record in their notebooks.

Dictation. Students could dictate specific information to an adult. The adult writes the information in the notebook for the student. Or the adult could write the sentence, using a highlighter, and students could trace the words, using a pencil.

Ownership

A student's science notebook can be personal or public. If the notebook is personal, the student decides how accessible his or her work is to other students. If ownership falls at the opposite extreme, everything is public, and anyone can look at the contents of anyone else's notebook at any time. In practice, most classroom cultures establish a middle ground in which a student's notebook is substantially personal, but the teacher claims free access to the student's work and can request that students share notebooks with one another and with the whole class from time to time.

Science Notebooks in Grades K–2

Notebook Components

- *Planning the investigation*
- *Data acquisition and organization*
- *Making sense of data*
- *Next-step strategies*

NOTEBOOK COMPONENTS

A few components give the science notebook conceptual shape and direction. These components don't prescribe a step-by-step procedure for how to prepare the notebook, but they do provide some overall guidance.

The general arc of an investigation starts with a question or challenge, and then proceeds with an activity, data acquisition, sense making, and next steps. The science notebook should record important observations and thoughts along the way. It may be useful to keep these four components in mind as you systematically guide students through their notebook entries.

Planning the Investigation

Typically at the start of a new activity, the first notebook entry is a focus question, which students glue or transcribe into their notebooks. The focus question determines the kinds of data to be collected and the procedures that will yield those data. Depending on the timing and their previous experiences, students may be asked to record a prediction related to the focus question.

Notebook entry with focus question from the Materials in Our World Module

Focus question. Each part of each investigation starts with a focus question or challenge. The focus question establishes the direction and conceptual challenge for the activity. Write or project it on the board or on the chart for students to transcribe into their notebooks, or give them photocopied strips of the focus question to tape or glue into their notebooks. The focus-question strips are distributed as needed. Some teachers give students adhesive-backed labels with the focus questions printed on them.

> **3. How can you sink wood?**
> Write the focus question on the chart, and have students read it together.
>
> ➤ *How can you sink wood?*

TEACHING NOTE

Student-developed plans appear infrequently in grades K–2. Asking students to record plans for every investigation is not recommended.

Plans and procedures. Students may plan their investigation. The planning may be detailed or informal, depending on the requirement of the investigation. These plans take time to develop and additional time to document in the notebook. A brief class discussion of the procedure may lead to a sentence or two recorded in the class notebook. This will be sufficient for many of the investigations.

Predictions. Depending on the content and the focus question, students may be able to make a prediction. When they make predictions, they are attempting to relate prior experiences to the question posed. Providing students with a frame can help them explain the rationale behind their predictions. A frame to help with stating a prediction is "I think that ________ because ________."

Data Acquisition and Organization

Data are the bits of information (observations) from which scientists construct ideas about the structure and behaviors of the natural world. Because observation is the starting point for answering the focus question, data records should be

- clearly related to the focus question;
- accurate and precise;
- organized for efficient reference.

Data handling can have two phases: data acquisition and data display. Data acquisition is making observations and recording data. The data record can be composed of words, phrases, numbers, and/or drawings. Data display is reorganizing the data in a logical way to facilitate thinking. The display can take the form of narratives, drawings, artifacts, tables, graphs, images, or other graphic organizers. Early in a student's experience with notebooks, the record may be disorganized and incomplete, and the display will need guidance. Students will need support to determine what form of recording to use in various situations and how best to display the data for analysis.

Narratives. The most intuitive approach to recording data for most students is narrative—using words, sentence fragments, and numbers in a more or less sequential manner. As students make a new observation, they record it right after the previous entry, followed by the next observation, and so on. Some observations, such as the changes observed in a mealworm, are appropriately recorded in narrative form.

Sentence frames provide a way for students to record their data. If students are nonwriters, they can dictate their observations, and you can write their observations in their notebooks, using a highlighter. Students can then trace the highlighted words.

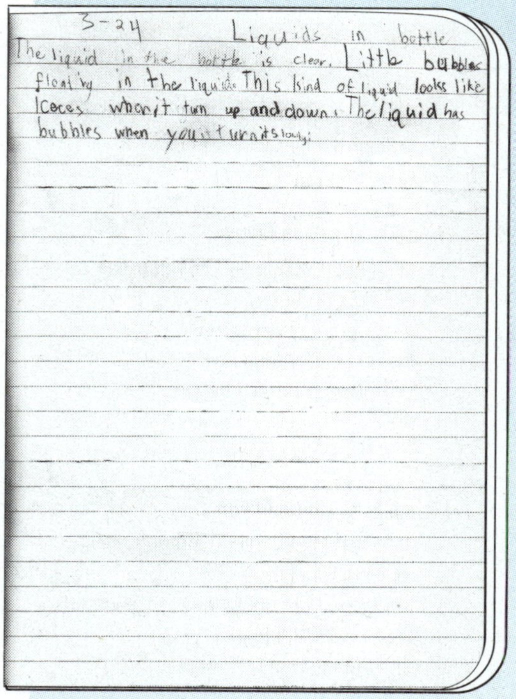

A narrative observation from the Solids and Liquids Module

Science Notebooks in Grades K–2

Drawings. When students observe organisms or systems, a labeled illustration is a very efficient way to record data. A picture is worth a thousand words, and a labeled picture is even more useful.

Some young students may initially prefer drawing their observations, while others may struggle to do so. When students make drawings, it can be helpful to suggest an acronym for making useful drawings. Accurate, big, colorful, and detailed (ABCD) drawings can capture structures of an organism, a balanced object, or an observation of a liquid. Think-alouds can also help students gain insight about not only what to draw but also when a drawing is useful.

It is not unusual for students to embellish their drawings by adding features such as smiles to flowers and living organisms. While this may be appropriate for creative expression, scientific illustrations should not be anthropomorphized. Give students feedback about making authentic observations.

Artifacts. Occasionally, the results of an investigation produce three-dimensional products that students can tape or glue directly into their science notebooks. Rubbings, disassembled fabrics, sand, minerals, seeds, and so on can become a permanent part of the record of learning.

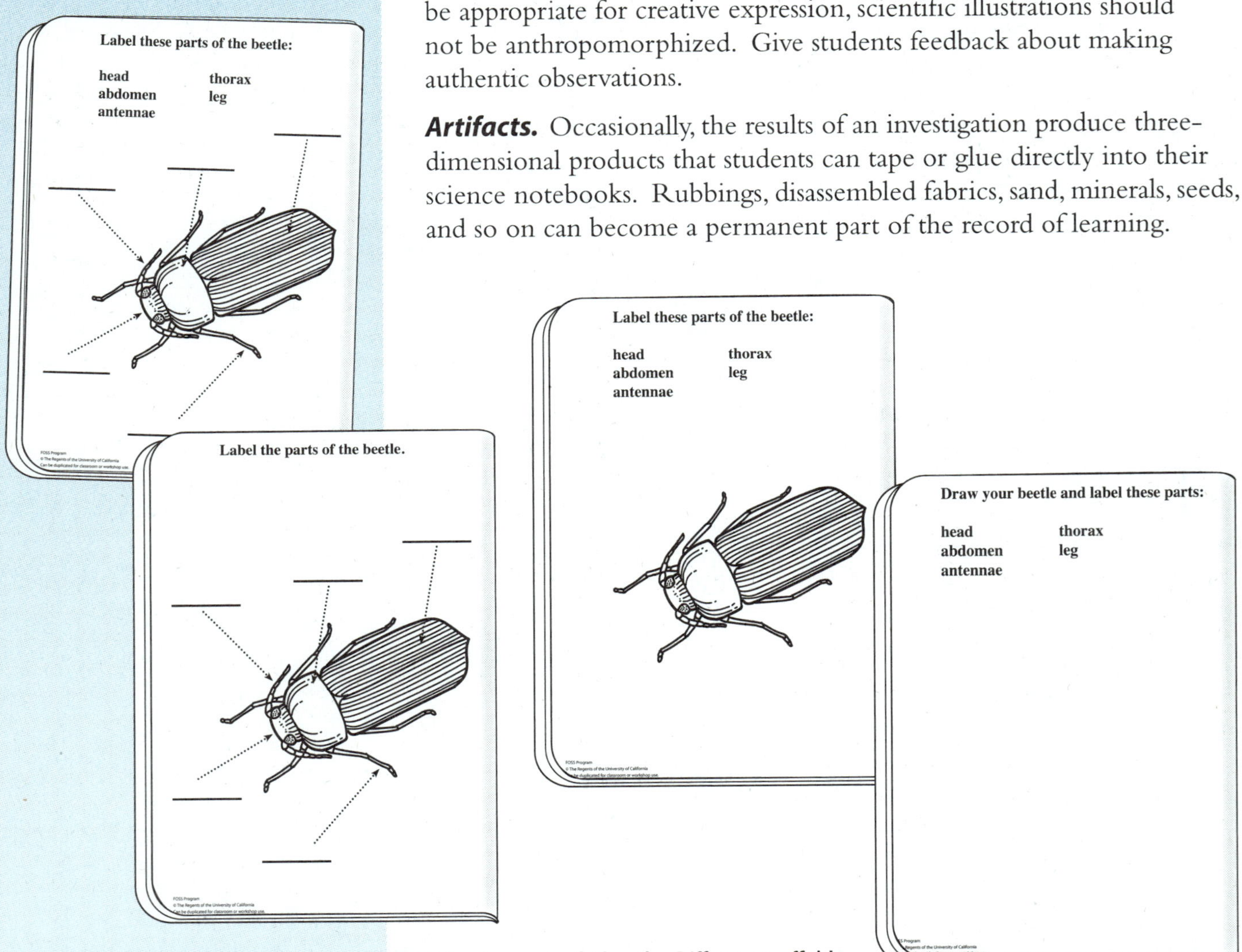

An example of different drawing scaffolds for the parts of a beetle. Different scaffolds can provide different levels of support for students. It is important to note that not all students need the same level of support.

Tables. When students make similar observations about a series of objects, such as properties of solid objects, a table with columns is an efficient recording method. The two-dimensional table makes it easy to compare the properties of all the objects under investigation.

Often, the preprinted notebook sheet provides a blank table. Students can quickly enter information on the notebook sheet. As students take on more independence in their notebooks, discussions about the column headings or the purpose of the tables shift the focus from filling in the table to the purpose the table serves.

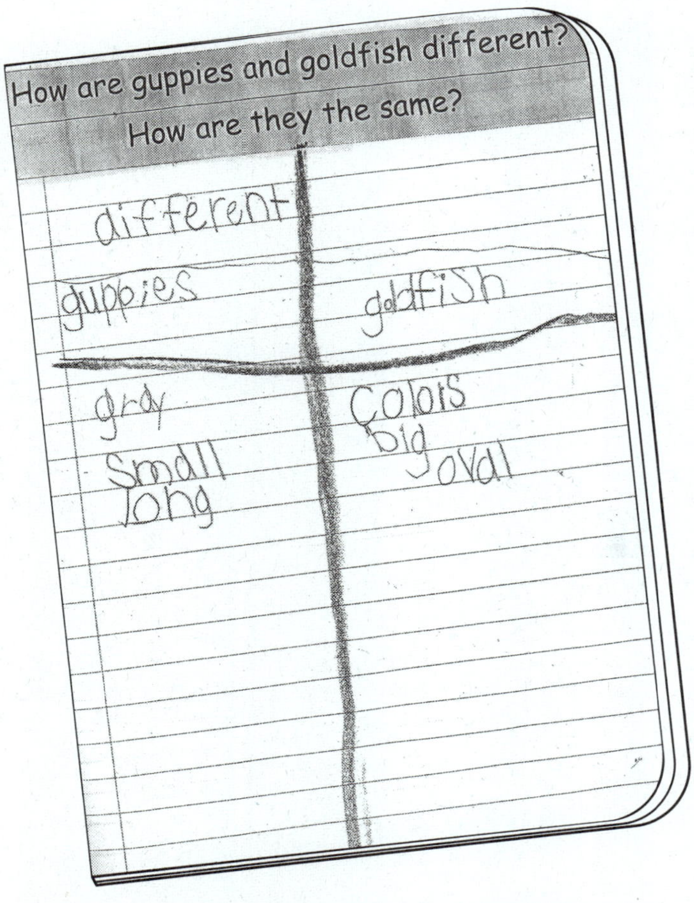

A simple table from the Animals Two by Two Module

Graphs and graphics. Reorganizing data into a logical, easy-to-use graphic is an important phase of data analysis. Bar graphs allow easy comparison. Additional graphic tools, such as concept maps and life cycles, help students make connections between data accrued during investigations.

For kindergarten students, you will need to model and scaffold graphs and graphics when students are ready. Visuals displayed in students' notebooks should also appear in a larger format on a whiteboard, projected display, or in the class notebook. You can use this larger display to connect organized data to the content students are learning.

Images. Digital photos of plants, rocks, and the results of investigations can be great additions to the science notebook.

Making Sense of Data

The third component of an investigation involves analyzing the data to learn something about the natural world. Establishing the habit of thinking about the data collected and using them to help us answer a question is important.

Most of the sense making in the primary classroom takes place during whole-group discussions when students share and discuss the observations they made while investigating. Questions are asked to help students interpret and analyze data in order to build conceptual understanding. Encourage students to use new vocabulary when they are sharing and making sense of their data. These words should be accessible to all students in the class notebook and displayed in the room. When students have limited written language skills, this oral discussion is important. This sharing is essential and is described in detail in the *Investigations Guide*. After sharing observations, the class revisits the focus question. Students flip back in their notebooks and use their data to discuss their answers to the question. You scaffold the discussion, and use appropriate language-development strategies.

8. Ask questions to guide discussion

Ask questions to guide students' observations and discussion.

- ➤ *Were you able to sink the wood by attaching paper clips?*
- ➤ *How many paper clips did you use to sink the wood?*
- ➤ *Let's compare the two samples. Does it take the same number of paper clips to sink both kinds of wood?*
- ➤ *Does it make a di!erence where you put the paper clips on the wood? All on one side? Evenly distributed around all sides?*

After the sense-making discussion, students might review relevant vocabulary. Then they answer the focus question in their notebooks. In the beginning, you might model this writing in a class notebook, or students might work in small groups to write a collaborative response. Students could also make a drawing as a response to a focus question. As students become more proficient writers, they will begin to record their own written responses to the focus question.

A sense-making entry from the Balance and Motion Module

Making sense of data is an opportunity for students to grapple with scientific concepts. The expectation is that all students will engage in this component. In many instances, using assistive structures, such as frames and prompts to guide the development of a coherent and complete response to the focus question, will help establish this expectation.

Thinking with evidence. In the primary classroom, students are expected to explain their thinking and provide some supporting evidence. While their explanations and evidence may be relatively simple, they will provide you with evidence of student understanding. A student might conclude that objects that have a round surface will roll because a ball and a can both rolled and they have round surfaces. The evidence should refer to specific observations, measurements, and so on.

TEACHING NOTE

If students respond verbally, it is important to record those responses in the science notebook to establish that notebooks are a place to document understandings.

For example, an investigation in the **Animals Two by Two Module** poses the focus question

➤ *What is the difference between red worms and night crawlers?*

Students observe and record, using the class notebook as a model. You create a Venn diagram, with input from students during discussion that compares the two types of worms. Students hear the focus question again, and they discuss their thinking in their groups. You write this frame in the class notebook

One difference between a red worm and a night crawler is ________.
I know this because ________.

All students get a preprinted frame to glue into their notebooks and answer independently. Be careful not to display conclusions or other sense-making entries in a class notebook before students make their own entries. Otherwise, assessment of student thinking becomes difficult, as a student may copy the class-notebook response with little understanding of the meaning.

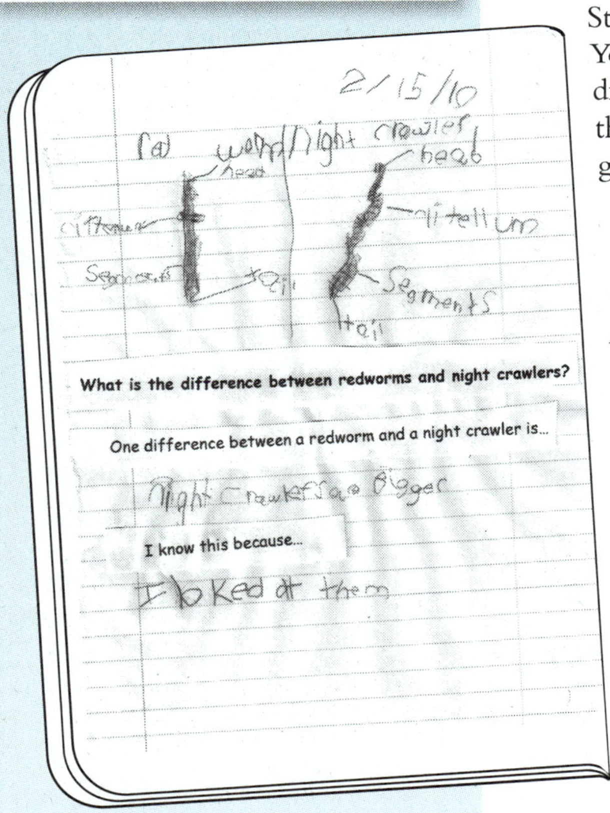

An answer to a focus question from the Animals Two by Two Module

Frames and prompts. Providing frames helps students organize their thinking. The frame provides a communication structure that allows students to focus on thinking about the science involved. This could be identifying the difference between two worms, students' thoughts about which types of objects roll, or how people can change the shape of wood. The frame does not do the thinking for students, but allows them to respond to the focus question in a clear, coherent manner.

- *One difference between the red worm and the night crawler is ________.*
- *I noticed that all the things that rolled ________.*
- *I know this because ________.*

TEACHING NOTE

Frames can be used to model how to write a good response. Some students may be able to revisit and modify previously used frames to answer additional questions.

I wonder. Does the investigation connect to a student's personal interests? Or does the outcome suggest a question or pique a student's curiosity? Providing time for students to write "I wonder" statements or questions supports the idea that the pursuit of scientific knowledge does not end with the day's investigation. The notebook is an excellent place to capture students' musings and for students to record thoughts that might otherwise be lost.

Wrap-up/warm-up. At the end of each investigation part or at the beginning of the next part, students will engage in a wrap-up or warm-up. This is another opportunity for students to revisit the content of the investigation. As they discuss their responses to the focus question with a partner, students may choose to edit their responses or add fresh information. This should be encouraged.

13. Share notebook entries

Conclude Part 4 or start Part 5 by having students share notebook entries. Ask students to open their science notebooks to the most recent entry. Read the focus question together as a class.

➤ *How can you sink wood?*

Ask students to pair up with a partner to

- share their answers to the focus question;
- explain their drawings.

TEACHING NOTE

Students' learning can be assessed only at the level in which it was done. If students worked in groups to answer the focus question, it is difficult to assess individual understanding. Similarly, providing a frame to guide a student response provides evidence on what students can do at a supported level, not at an independent level.

TEACHING NOTE

These next-step strategies should be kept simple for primary students. The key idea is that students can revise their responses after gaining more information.

Next-Step Strategies

In each investigation, the *Investigations Guide* indicates an assessment opportunity and what to look for when examining students' work. The purpose of looking at students' work at this juncture is for formative, or embedded, assessment, *not* for grading. Look for patterns in students' understanding by collecting and sorting the notebooks. If the patterns indicate that students need additional help with communication or with content, you might want to select a next-step strategy before going on to the next part. This process of looking at students' work is described in more detail in the Assessment chapter.

A next-step strategy is an instructional tool designed to help students clarify their thinking and usually takes place before the start of the next investigation part. A strategy is selected based on students' needs. It may be that a student needs to communicate his or her thinking more efficiently or accurately or to use scientific vocabulary. A student may need to think about the concept in a different way. Because many young students are not able to articulate their thinking well in writing, it can be difficult to discern the area of need.

What follows is a collection of next-step strategies that teachers have used successfully with groups of students to address areas of need. These strategies are flexible enough to use in different groupings and can be modified to meet your students' needs.

Teacher feedback. Students' writing often exposes weaknesses in students' understanding—or so it appears. It is important to check whether the flaw results from poor understanding of the science or from imprecise communication. When students are able to read your comments, you can use self-stick notes to provide constructive feedback or dig deeper into students' thinking. The note might pose a question designed to move a student's thinking forward or to clarify an explanation. The student acts independently on the question.

When students have not developed the reading skills necessary to act independently on written feedback, read the question or prompt to the student. Language skills are supported when students have the opportunity to connect the written feedback to what you read to them.

The most effective forms of feedback relate to the content of the work. Here are a couple of examples.

- *Label your drawing to show the parts of the plant.*
- *Tell me why you think you were able to balance the triangle and the arch.*

Nonspecific feedback (such as stars, smiley faces, and "good job!") and ambiguous critiques, (such as "try again," "put more thought into this," and "not enough,") are less effective. Feedback that guides students to think about the content of their work and gives suggestions for how to improve are productive instructional strategies. Here are some examples of useful generic feedback.

- *Use the science words in your answer.*
- *Can you tell me why you think that?*
- *Why do you think that happened?*

Students return to their notebooks and read or listen to the feedback at the start of the next investigation. They can discuss this feedback with a partner during the warm-up time and refine their responses. You could model this refinement as a think-aloud or by using the class notebook. Monitor students to ensure that they are acting on the feedback provided.

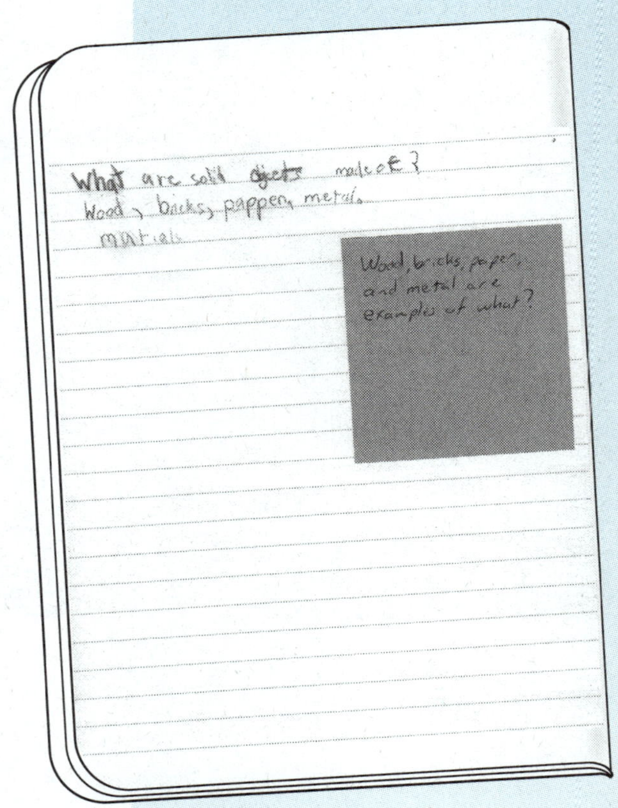

Students can respond to feedback by using a different color to show their new understanding.

Review and critique anonymous student work. Presenting work from other students can be a valuable learning tool for refining and improving the content and literacy of responses. Depending upon the culture of the class, you might present actual or simulated student work from a focus question or responses that reflect a common misconception, error, or exemplary work. Present it to the class in a common gathering area, such as on a rug, or display it electronically while students are at their seats. Present one simple response, such as a labeled drawing. Students then work in a group to discuss the merits and recommend improvements to the student response. In this process, students discuss what information is needed for a quality response. After critiquing other students' responses, students look at their own responses and refine their thinking. Students could also use a different-color crayon or pencil to make changes to their own responses.

Key points. Pose the focus question to the class, and, through discussion, elicit the key ideas or points that would completely answer the question. List only key words or brief phrases on the board. If you have already recorded the key points in the class notebook, revisit that list as well. Once the class has agreed on the key points, students review their own responses, looking for the key points. You can guide this by calling out each key point on the list, and asking students to put a finger on it in their notebooks. If students need to add the key point, they can add it in another color.

Mini-lessons. Sometimes the data from sorting notebook entries reveal that students need some information repeated or specific guidance on a skill. A mini-lesson is a brief interaction with a group of students that addresses a specific area of need. You might have a group of students observe a mealworm more closely in order to count the number of segments, or give students a writing prompt and work with them to explain their thinking more clearly.

WRITING OUTDOORS

Every time you go outdoors with students, you will have a slightly different experience. Naturally, the activity or task will be different, but other variables may change as well. The temperature, cloud cover, precipitation, moisture on the ground; other activities unexpectedly happening outside; students' comfort levels related to learning outdoors; and time of the school year are all aspects that could affect the activity and will certainly determine how you incorporate the use of notebooks. The following techniques are tried-and-true ways to help students learn how to write outdoors and to give them all the supplies they need to support their writing.

Create "Desks"

Students need a firm writing surface. Students who write in composition notebooks with firm covers can simply fold them open to the pages they are writing on, rest them in the crook of their nonwriting arms, and hold them steady with their nonwriting hands—they can stand, sit, kneel, or lean against a wall to write. At the beginning of the school year, take a minute to model how to do this.

Many students feel most comfortable sitting down to write. Curbs, steps, wooden stumps or logs, rocks, and grass are places to sit while writing. Select a writing location that suits your students' comfort levels. Some students will not be comfortable sitting on the grass or ground at first. They will need to sit on something such as a curb, boulder, or wooden stump at the start of the year, but will eventually feel more comfortable with all aspects of the outdoor setting as the year moves along.

If students are using individual notebook sheets or notebooks with flimsy covers, you will likely want to buy or make clipboards. If you do not have clipboards, use a box cutter to cut cardboard to the proper size. Clamp a binder clip at the top to make a lightweight yet sturdy clipboard. If it gets ruined, no tears will be shed. If you're in the market for new clipboards, get the kind that are stackable and do not have a bulky clip. Ideally, all the clipboards will fit in one bag for portability and easy distribution.

If using a notebook sheet, simply put the sheet on the clipboard before going outdoors, and have students glue the sheet into their notebooks when you are back in the classroom. An elastic band around the bottom of the clipboard, or around a stiffer composition notebook, will help keep the paper from flapping around and becoming too weathered.

Bring Writing Tools Outdoors

Almost always, students will set up their notebooks indoors so that they know what is expected of them outdoors. If students will be recording data, they will carry their pencils and notebooks or clipboards outdoors and hold on to them the entire time. Make sure students understand how and what to record.

You can bring chart paper outdoors. Roll up a blank piece of chart paper, grab some blue painter's tape, and stick the piece of paper to the school wall. You'll need to tape all four corners. Or set up a chart inside and clip it to a chain-link fence with binder clips or clothespins when you go outdoors.

Take along extra pencils, as many pencil points will break. Some teachers find it helpful to tie pencils onto clipboards. Pencils should be tucked between the clip and the notebook sheet so that students don't poke themselves or, more likely, accidentally break the pencil points.

Another option is to prepare a cloth bag containing small pencil sharpeners, extra pencils, and other science tools, such as vials, hand lenses, and rulers.

Decide When to Write Outdoors

In general, notebook entries will be more detailed and more insightful if students can stay outdoors where the scientific exploration occurred. Sometimes, you will want to complete notebook entries indoors. If you are teaching the module early in the year when students are building up the routines for using the schoolyard, or if the weather is not ideal (a little chilly, raining, too hot, too windy), then you may want students to make notebook entries after returning to the classroom. If students are totally focused and in the moment, they can stay outdoors while they answer the focus question. If other students are outdoors playing, you may need to bring the class indoors to complete the written work. Only you will be able to determine what is best at the time.

CLOSING THOUGHTS

Engaging primary students in active science with notebooks provides a rich experience. Doing this successfully requires thoughtful interactions among students, materials, and natural phenomena. Initially, adding notebooks to your science teaching will require you to focus students' attention on how to set up the notebook, what types of entries students should make, and when students should be using their notebooks. You will establish conventions about where to record the date and title, where to keep notebooks, how to glue notebook sheets into notebooks, and when to record observations and thinking.

Once you are past these perfunctory issues, you can shift your focus to the amount of scaffolding to provide to students or to encouraging students to create their own notebook entries. During this time, you and your students are developing skills to improve the quality of notebook entries. These skills may include asking better questions to focus students' attention on a specific part of an organism or using color to enhance a drawing. Students begin to make entries with less prompting. They give more thought to supporting their responses to the focus question. When asked to make a derivative product, students thumb through their notebooks to find the needed information. The notebook becomes a tool for students to help recall their learning.

NOTE
For more on derivative products, see the Science-Centered Language Development chapter.

As students begin to document their thinking about focus questions and other queries, you may begin to wonder, "Should I be doing something with their notebooks?" This is when your focus shifts from the notebook as just something students use during science learning to the notebook as an assessment tool. Once everyone is comfortable recording the focus question and collecting data, you can take the next step of collecting notebooks and reading students' responses as a measure of not just how individual students are learning, but what the pervasive needs of students are. You choose next-step strategies that address students' needs before proceeding to the next investigation. The notebooks act as an assessment tool that lets you modify your science instruction.

This process will take time, discussions with colleagues, revisiting different sections of this chapter, and critical scrutiny of students' work before both you and your students are using notebooks to their full potential.

Science-Centered Language Development

Science-Centered Language Development

Teams of science inquirers talk about and write about their questions, their tentative explanations, their relationships between evidence and explanations, and their reasons and judgments about public presentations and scientific arguments in behalf of their work. It is in the context of this kind of scientific activity that students' literacy of the spoken and written word develops along with the literacy of the phenomenon.

Hubert M. Dyasi, "Visions of Inquiry: Science"

Contents

INTRODUCTION

In this chapter, we explore the intersection of science and language and the implications for effective science teaching and language development. We identify best practices in language arts instruction that support science learning and examine how learning science content and practices supports language development. The active investigations, science notebooks, *FOSS Science Resources* readings, and formative assessment activities in FOSS provide rich contexts in which students develop and exercise thinking processes and communication skills. Together, these elements comprise effective instruction in both science and language arts—students experience the natural world around them in real and authentic ways and use language to inquire, process information, and communicate their thinking about scientific phenomena. We refer to the development of language within the context of science as science-centered language development.

Science-Centered Language Development

Language plays two crucial roles in science learning: (1) it facilitates the communication of conceptual and procedural knowledge, questions, and propositions (external, public), and (2) it mediates thinking, a process necessary for understanding (internal, private). These are also the ways scientists use language: to communicate with one another about their inquiries, procedures, and understandings; to transform their observations into ideas; and to create meaning and new ideas from their work and the work of others. For students, language development is intimately involved in their learning about the natural world. Science provides a real and engaging context for developing literacy, and language arts skills and strategies support conceptual development and scientific practices. For example, the skills and strategies used for reading comprehension, writing expository text, and oral discourse are applied when students are recording their observations, making sense of science content, and communicating their ideas. Students' use of language improves when they discuss, write, and read about the concepts explored in each investigation.

We begin our exploration of science and language by focusing on language functions and how specific language functions are used in science to facilitate information acquisition and processing (thinking). Then we address issues related to the specific language domains—speaking and listening, writing, and reading. Each section addresses

- how skills in that domain are developed and exercised in FOSS science investigations;
- literacy strategies that are integrated purposefully into the FOSS investigations;
- suggestions for additional literacy strategies that both enhance student learning in science and develop or exercise English-language literacy skills.

Following the domain discussions is a section on science-vocabulary development, with scaffolding strategies for supporting all learners. The last section covers language-development strategies specifically for English learners.

NOTE

The term *English learners* refers to students who are learning to understand English. This includes students who speak English as a second language and native English speakers who need additional support to use language effectively.

THE ROLE OF LANGUAGE IN SCIENTIFIC AND ENGINEERING PRACTICES

Language functions are the purpose for which speech or writing is used and involve both vocabulary and grammatical structure. Understanding and using language functions appropriately is important in effective communication. Students use numerous language functions in all disciplines to mediate communication and facilitate thinking (e.g., they plan, compare, discuss, apply, design, draw, and provide evidence).

In science, language functions facilitate the enactment of scientific and engineering practices. For example, when students are *collecting data*, they are using language functions to identify, label, enumerate, compare, estimate, and measure. When students are *constructing explanations*, they are using language functions to analyze, communicate, discuss, evaluate, and justify.

A Framework for K–12 Science Education: Practices, Crosscutting Concepts, and Core Ideas (National Research Council 2012) provides a "set of essential practices as essential for any education in the sciences and engineering." Each of these scientific and engineering practices requires the use of multiple language functions. Often, these language functions are part of an internal dialogue weighing the merits of various explanations—what we call thinking. The more language functions with which we are facile, the more effective and creative our thinking can be.

The scientific and engineering practices are listed below, along with a sample of the language functions that are exercised when effectively engaged in that practice. (Practices are bold; language functions are italic.)

Asking questions and defining problems

- *Ask* questions about objects, organisms, systems, and events in the natural world (science) or define a problem (engineering).

Planning and carrying out investigations; analyzing, and interpreting data

- *Plan* and conduct investigations.
- *Observe* and *record* data.
- *Measure* to extend the senses to acquire data.
- *Organize* observations (data), using numbers, words, images, and graphics.

Examples of Language Functions

Analyze
Apply
Ask questions
Clarify
Classify
Communicate
Compare
Conclude
Construct
Critique
Describe
Design
Develop
Discuss
Distinguish
Draw
Enumerate
Estimate
Evaluate
Experiment
Explain
Formulate
Generalize
Group
Identify
Infer
Interpret
Justify
Label
List
Make a claim
Measure
Model
Observe
Organize
Plan
Predict
Provide evidence
Reason
Record
Represent
Revise
Sequence
Solve
Sort
Strategize
Summarize

Constructing explanations (science) and designing solutions (engineering); engaging in argument from evidence

- *Predict* future events, based on evidence and reasonings.
- Use data and logic to *construct* and *communicate* reasonable explanations.
- *Develop*, *discuss*, *evaluate*, and *justify* the merits of explanations.
- Use science-related technologies to *apply* scientific knowledge and *solve* problems.

Developing and using models

- Create models to *explain* natural phenomena and *predict* future events or outcomes.
- *Describe* how models represent the natural world.
- *Compare* models to actual phenomena and *identify* limitations of the models.

Obtaining, evaluating, and communicating

- *Identify* key ideas in text, identify supporting evidence, and know how to use various text features.
- *Distinguish* opinion from evidence.
- Use writing as a tool for *clarifying* ideas and *communicating*.

Research supports the claim that when students are intentionally using language functions in thinking about and communicating in science, they improve not only science content knowledge, but also language-arts and mathematics skills (Ostlund 1998; Lieberman and Hoody 1998). Language functions play a central role in science as a key cognitive tool for developing higher-order thinking and problem-solving abilities that, in turn, support academic literacy in all subject areas.

Here is an example of how an experienced teacher can provide an opportunity for students to exercise language functions in FOSS. In the **Soils, Rocks, and Landforms Module**, this is one piece of content we expect students to have acquired by the end of the module.

- The greater the flow of water across Earth's surface, the greater the rate of erosion and deposition.

The scientific practices the teacher wants the class to focus on are *interpreting data* and *constructing explanations*.

The language functions students will exercise while engaging in these scientific practices are *comparing*, *explaining*, and *providing evidence*. The teacher understands that these language functions are appropriate to the purpose of the science investigation and support the Common Core Standards for writing (students will write explanatory texts to examine a topic and convey ideas and information clearly).

- Students will *compare* observational data (from a stream-table investigation) to *explain* the relationship between the amount of water that runs over a surface and the amount of erosion and deposition that occur.

The teacher can support the use of language functions by providing structures such as sentence frames.

- As _____, then _____.

  As more water flows over a surface, then more erosion and deposition occur.

- When I changed _____, then _____ happened.

  When I changed the amount of water flowing down the stream table, then more erosion and deposition happened.

- The more/less _____, the _____.

  The more water that flows across the earth materials, the more erosion and deposition that occur.

NOTE

For more examples of how FOSS teachers address language-arts standards while conducting science investigations with students, go to FOSSweb (www.FOSSweb.com).

SPEAKING AND LISTENING DOMAIN

The FOSS investigations are designed to engage students in productive oral discourse. Talking requires students to process and organize what they are learning. Listening to and evaluating peers' ideas calls on students to apply their knowledge and to sharpen their reasoning skills. Guiding students in small-group and whole-class discussions is critical to the development of conceptual understanding of the science content and the ability to think and reason scientifically.

FOSS investigations start with a discussion—either a review to activate prior knowledge or presentation of a focus question or a challenge to motivate and engage active thinking. During active investigation, students talk with one another in small groups, share their observations and discoveries, point out connections, ask questions, and start to build explanations. The discussion icon in the sidebar of the *Investigations Guide* indicates when small-group discussions should take place.

During the activity, the *Investigations Guide* indicates where it is appropriate to pause for whole-class discussions to guide conceptual understanding. The *Investigations Guide* provides you with discussion questions to help stimulate student thinking and support sense making. At times, it may be beneficial to use sentence frames or standard prompts to scaffold the use of effective language functions and structures.

At the end of the investigation, there is another opportunity to develop language through speaking and listening in the Wrap-Up/Warm-Up. During this time, students are often asked to discuss their responses to the focus question. All students need to be engaged in this opportunity to use different language functions.

On the following pages are some suggestions for providing structure to those discussions and for scaffolding productive discourse when needed. Teaching techniques used to generate discussion in language arts and other content areas can also be used effectively during science. Using the protocols that follow will ensure inclusion of all students in discussions.

FOSS

Partner and Small-Group Discussion Protocols

Whenever possible, give students time to talk with a partner or in a small group before conducting a whole-class discussion. This provides all students with a chance to formulate their thinking, express their ideas, practice using the appropriate science vocabulary, and receive input from peers. Listening to others communicate different ways of thinking about the same information from a variety of perspectives helps students negotiate the difficult path of sense making for themselves.

Partner and Small-Group Discussion Protocols
- *Dyads*
- *Partner parade*
- *Put in your two cents*

Dyads. Students pair up and take turns either answering a question or expressing an idea. Each student has 1 minute to talk while the other student listens. While student A is talking, student B practices attentive listening. Student B makes eye contact with student A, but cannot respond verbally. After 1 minute, the roles reverse.

Here's an example from the **Mixtures and Solutions Module**. Just before students answer the focus question in their notebooks, you ask students to pair up and take turns sharing their answer to the question "How do you know when a chemical reaction has occurred?" The language objective is for students to be able to explain what happens during a chemical reaction and provide evidence (orally and in writing) that it has occurred. These sentence frames can be written on the board to scaffold student thinking and conversation.

- I claim that ______ produces a chemical reaction, because ______.
- Evidence for my claim includes ______.

Partner parade. Students form two lines facing each other. Present a question, an idea, an object, or an image as a prompt for students to discuss. Give students 1 minute to greet the person in front of them and discuss the prompt. After 1 minute, call time. Have the first student in one of the lines move to the end of the line, and have the rest of the students in that line shift one step sideways so that everyone has a new partner. (Students in the other line do not move.) Give students a new prompt to discuss for 1 minute with their new partners.

For example, students are just beginning the investigation on landforms, and you want to assess prior knowledge. Give each student a picture of a landform, and have students line up facing each other. For the first round, ask, "What do you know about the image on your card?" For the second round, ask, "How do you think the landform in your picture formed?" For the third round, ask, "What personal experience have you had with this landform?" The language objective is for students to describe their observations, infer how the landform formed, and reflect upon and relate any experiences they may have had with a similar landform. The following sentence frames can be used to scaffold student discussion.

- I notice ____.
- I think the landform was formed by ____.
- The landform in my picture reminds me of the time when ____.

Put in your two cents. For small-group discussions, give each student two pennies or similar objects to use as talking tokens. Each student takes a turn putting a penny in the center of the table and sharing his or her idea. Once all have shared, each student takes a turn putting in the other penny and responding to what others in the group have said. For example,

- I agree (or don't agree) with ____ because ____.

Here's an example from the **Sun, Moon, and Planets Module**. Students have been monitoring their shadows throughout the day and are still struggling to make coherent connections between Earth's rotation and the effect that it has on shadows. The language objective is for students to describe their observations, explain why the direction and length of their shadows change during the day, and provide evidence based on their own observations. You give each student two pennies, and in groups of four, they take turns putting in their two cents. For the first round, each student answers the question "Why do shadows change during the day?" They use the frame

- I think shadows change during the day because ____.
- My evidence is ____ .

On the second round, each student states whether he or she agrees or disagrees with someone else in the group and why, using the sentence frame.

Whole-Class Discussion Supports
- *Sentence frames*
- *Guiding questions*

Whole-Class Discussion Supports

The whole-class discussion is a critical part of sense making. After students have had the active learning experience and have talked with their peers in partners and/or small groups, sharing their observations with the whole class sets the stage for developing conventional explanatory models. Discrepant events, differing results, and other surprises are discussed, analyzed, and resolved. It is important that students realize that science is a process of finding out about the world around them. This is done through asking questions, testing ideas, forming explanations, and subjecting those explanations to logical scrutiny. Leading students through productive discussion helps them connect their observations and the abstract symbols (words) that represent and explain those observations. Whole-class discussion also provides an opportunity for you to interject an accurate and precise verbal summary statement as a model of the kind of thinking you are seeking. Facilitating effective whole-class discussions takes skill, practice, a shared set of norms, and patience. In the long run, students will have a better grasp of the content and improve their abilities to think independently and communicate effectively.

Norms should be established so that students know what is expected during science discussions.

- Science content and practices are the focus.
- Everyone participates.
- Ideas and experiences are shared, accepted, and valued. Everyone is respectful of one another.
- Claims are supported by evidence.
- Challenges (debate and argument) are part of the quest for complete understanding.

TEACHING NOTE

Let students know that scientists change their minds based on new evidence. It is expected that students will revise their thinking based on evidence presented in discussions.

The same discussion techniques used during literacy circles and other whole-class discussions can be used during science instruction (e.g., attentive listening, staying focused on the speaker, asking questions, responding appropriately). In addition, in order for students to develop and practice their reasoning skills, they need to know the language forms and structures and the behaviors used in evidence-based debate and argument (e.g., using data to support claims, disagreeing respectfully, asking probing questions; Winokur and Worth 2006).

Explicitly model and conduct mini-lessons (5 to 10 minutes of focused instruction) on the language structures appropriate for active discussions, and provide time for students to practice them, using the science content.

TEACHING NOTE

Encourage science talk. Allow time for students to engage in discussions that build on other students' observations and reasoning. After an investigation, use a teacher- or student-generated question, and either just listen or facilitate the interaction with questions to encourage expression of ideas among students.

Sentence frames. The following samples can be posted as a scaffold as students learn and practice their reasoning and oral participation skills.

- I think ______, because ______.
- I predict ______, because ______.
- I claim ______; my evidence is ______.
- I agree with ______ that ______.
- My idea is similar/related to ______'s idea.

- I learned/discovered/heard that _____.
- <Name> explained _____ to me.
- <Name> shared _____ with me.
- We decided/agreed that _____.
- Our group sees it differently, because _____.
- We have different observations/results. Some of us found that _____. One group member thinks that _____.
- We had a different approach/idea/solution/answer _____.

Guiding questions. The *Investigations Guide* provides questions to help concentrate student thinking on the concepts introduced in the investigation. Guiding questions should be used during the whole-class discussion to facilitate sense making. Here are some other open-ended questions that help guide student thinking and promote discussion.

- What did you notice when _____?
- What do you think will happen if _____?
- How might you explain _____? What is your evidence?
- What connections can you make between _____ and _____?

Whole-Class Discussion Protocols

The following examples of tried-and-true participation protocols can be used to enhance whole-class discussions during science and all other curriculum areas. The purpose of these protocols is to increase meaningful participation by giving all students access to the discussion, allowing students time to think (process), and providing a context for motivation and engagement.

Think-pair-share. When asking for a response to a question posed to the class, allow time for students to think silently for a minute. Then, have students pair up with a partner to exchange thoughts before you call on a student to share his or her ideas with the whole class.

Pick a stick. Write each student's name on a craft stick, and keep the sticks handy at the front of the room. When asking for responses, randomly pick a stick, and call on that student to start the discussion. Continue to select sticks as you continue the discussion. Your name can also be on a stick in the cup. You can put the selected sticks in a different location or back into the same cup to be selected again.

Whip around. Each student takes a quick turn sharing a thought or reaction. Questions are phrased to elicit quick responses that can be expressed in one to five words (e.g., "Give an example of a stored-energy source." "What does the word *heat* make you think of?").

Group posters. Have small groups design and graphically record their investigation data and conclusions on a quickly generated poster to share with the whole class.

Whole-Class Discussion Protocols

- *Think-pair-share*
- *Pick a stick*
- *Whip around*
- *Group posters*

Two-cup pick-a-stick container

One-cup pick-a-stick container divided with tape

WRITING DOMAIN

Information processing is enhanced when students engage in informal writing. When allowed to write expressively without fear of being scorned for incorrect spelling or grammar, students are more apt to organize and express their thoughts in different ways that support sense making. Writing promotes the use of scientific practices, thereby developing a deeper engagement with the science content. It also provides guidance for more formal derivative science writing (Keys 1999).

NOTE
For more information about supporting science-notebook development, see the Science Notebooks chapter.

Science Notebooks

The science notebook is an effective tool for enhancing learning in science and exercising various forms of writing. Science notebooks provide opportunities both for expressive writing (students craft explanatory narratives that make sense of their science experiences) and for practicing informal technical writing (students use organizational structures and writing conventions). Starting as emergent writers in kindergarten, students learn to communicate their thinking in an organized fashion while engaging in the cognitive processes required to develop concepts and build explanations. Having this developmental record of learning also provides an authentic means for assessing students' progress in both scientific thinking and communication skills.

NOTE
Language forms and structures refer to the internal grammatical structure of words and how those words go together to make sentences.

One way to help students develop the writing skills necessary for productive notebook entries is to focus on the corresponding language functions. The language forms and structures used to perform these language functions in science are used in all curricular areas and, therefore, can be suitably taught in conjunction with existing language-arts instruction. This can be done through mini-lessons on the writing skills that support the various types of notebook entries.

Table 1, at the end of the Writing Domain section, provides examples of how language functions are used to help students develop both their general writing skills and their thinking abilities within the format of the science-notebook entry. The writing objectives for a mini-lesson, along with the particular language forms and structures (vocabulary, syntax, linking words, organization of ideas, and so on), are identified in the table along with the suggested sentence frames for scaffolding.

For example, if students are observing a particular insect's behavior, the language objective might be "Students describe in detail the behavior of the insect." A prior mini-lesson on using sensory details to describe observations would provide students with the language forms and structures appropriate for recording their data in their notebooks during the observations. As a scaffold, students could also be provided with sentence frames to help them write detailed narratives.

Language function	Language objectives for writing in notebooks	Language forms, structures, and scaffolds for writing
Notebook component—data acquisition		
Describe	Write narratives: use details, sensory observations, and connections to prior knowledge.	I observed ___. When I touch the ___, I feel ___. The ___ has ___. I noticed ___. It feels ___. It smells ___. It sounds ___. It reminds me of ___, because ___.

This sample from Table 1 shows how language functions can be developed and applied when writing in science notebooks.

Developing Derivative Language-Arts Products

Science notebooks provide students with a source of information (content) from which they can draw to create more formal science-centered writing pieces. Derivative products are written pieces that are generated with specific language-arts goals in mind, such as audience and language function. Writing-to-learn methods can enhance science concepts when students engage in different types of writing for different purposes (Hand and Prain 2002). We know that students are more engaged and motivated to write when they have a clear and authentic context for writing.

The language extensions in the Interdisciplinary Extensions section at the end of each investigation suggest writing activities that can be used to help students learn the science content for that particular investigation. The writing activities incorporate language-arts skills appropriate for the grade level. Questions and ideas for future writing activities that surface during the investigations can be recorded in a class list, in science notebooks, or in students' writing folders. Here are general suggestions for using science content to create products in each of several writing genres.

Descriptive writing. Students use descriptive writing to portray an organism, an environment, an object, or a phenomenon in such a way that the reader can almost recapture the writer's experience. This is done through the use of sensory language; rich, vivid, and lively detail; and figurative language, such as simile, hyperbole, metaphor, and symbolism. You can remind students to show, rather than tell, through the use of active verbs and precise modifiers.

To help them learn the science content, students can use the information in their science notebooks to elaborate on their observations by using descriptive vocabulary, analogies, and drawing.

NOTE
The complete table appears at the end of this Writing Domain section.

Derivative Language-Arts Products

- *Descriptive writing*
- *Persuasive writing*
- *Narrative writing*
- *Expository writing*
- *Recursive cycle*

For example, kindergartners can draw pictures of isopods, fish, worms, things made of wood, different types of trees and leaves, and so on. Through drawings and words, students describe objects' properties and compare how those objects are the same and different.

Primary students can make property cards by writing on an index card as many properties as they can that describe an object or organism. Then, students take turns reading the properties to another student to see if the partner can identify the corresponding object.

Students of all ages can expand on their notebook entries in poetry that expresses their interpretation of organisms, objects, and phenomena such as crayfish, seeds, electricity, rivers, weather, and minerals. The use of similes is a good way to engage students in making comparisons. You can use a simple frame, such as: ______ is like ______. (For example, in the **Air and Weather Module**, students might write "The weather today is like a dragon.")

Students can share their similes. Other students can explain why they think the simile works.

Persuasive writing. The objective of persuasive writing is to convince the reader that a stated opinion or interpretation of data is worthwhile and meaningful. Students learn to support their claims with evidence and to use persuasive techniques, such as logical arguments and calls to action. By using claims and evidence to formulate conclusions in their science notebooks, students develop and apply their thinking and reasoning skills to form the basis of persuasive writing in a variety of formats, such as essays, letters, editorials, advertisements, award nominations, informational pamphlets, and petitions. Animal habitats, energy use, weather patterns, landforms, and water sources are just a few science topics that can generate questions and issues for persuasive writing.

Here is a sample of persuasive writing frames (modified from Gibbons 2002).

Title: ____
The topic of this discussion is ____.
My opinion (position, conclusion) is ____.
There are <number> reasons why I believe this to be true.
First, ____.
Second, ____.
Finally, ____.
On the other hand, some people think ____.
I have also heard people say ____.
However, my claim is that ____ because ____.

Narrative writing. Science provides a broad landscape of engaging material for stimulating the imagination for the writing of stories, songs, biographies, autobiographies, poems, and plays. Students can use organisms or objects as characters; describe habitats and environments as settings; and write scripts portraying various systems, such as weather patterns, flow of electricity, and water, rock, or life cycles.

Expository writing. Students use science content to inform, explain, clarify, define, or instruct through writing letters, definitions, procedures, newspaper and magazine articles, posters, pamphlets, and research reports. Expository writing is characterized by a focus on main topics with supporting facts, details, explanations, and examples and is organized in a clear, coherent, and sequential manner. Expository writing has a clear focus on communicating accurate, complete, and detailed representations of science content and/or observation of natural history or events.

During writing instruction, students can use the information in their science notebooks and in related readings (and other sources, such as video content) to write a more formal and conclusive answer to the focus question. Strategies such as the writing process (plan, draft, edit, revise, and share) and writing frames (modeling and guiding the use of topic sentences, transition and sequencing words, examples, explanations, and conclusions) can be used with the science content to develop proficiency in critical writing skills.

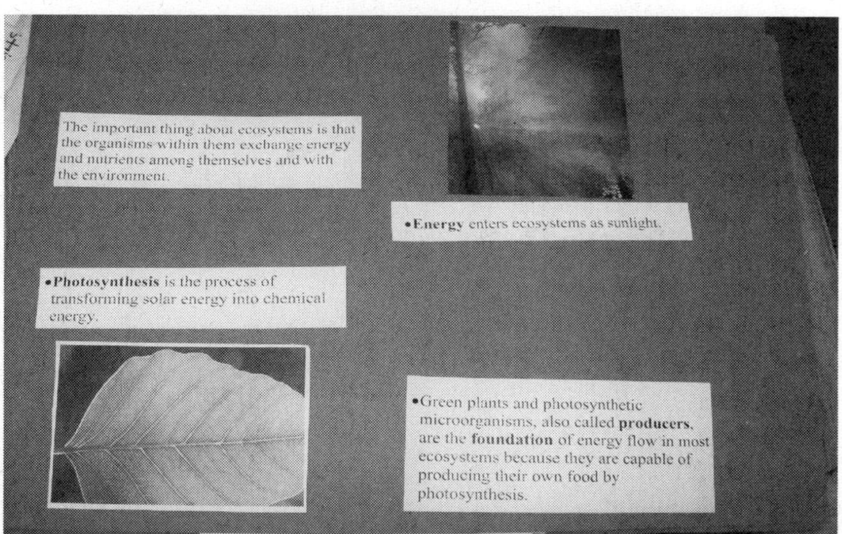

A page from a big book about ecosystems, a derivative product created during language arts

▶NOTE
Human characteristics should not be given to organisms (anthropomorphism) in science investigations, only in literacy extensions.

Here are three samples of writing frames for science reports for primary and upper-elementary students (modified from Wellington and Osborne 2001).

Primary (life cycle appropriate for the **Insects and Plants Module**)

Title: _____

(Define) A _____ is _____.

(Classify) A _____ is a kind of _____.

(Describe) A _____ is _____.

(Habitat) _____ lives _____.

(Life cycle) _____ started life as _____.

(Change over time) Then it _____.

(Food) A _____ eats _____.

(Evidence) I know that _____ because _____.

Upper elementary (living structures appropriate for the **Structures of Life Module**)

Title: _____

(Identify) The part of the body I am describing is the _____.

(Describe) It consists of _____.

(Explain) The function of these parts is _____.

(Example) This drawing shows _____.

Upper elementary (explanation)

Title: _____

I want to explain why (how) _____.

An important reason for why (how) this happens is that _____.

Another reason is that _____.

I know this because _____.

Recursive cycle. An effective method for extending students' science learning through writing is the recursive cycle of research (Bereiter 2002). This strategy emphasizes writing as a process for learning, similar to the way students learn during the active science investigations.

1. Decide on a problem or question to write about.
2. Formulate an idea or a conjecture about the problem or question.
3. Identify a remedy or an answer, and develop a coherent discussion.
4. Gather information (from science notebooks, *FOSS Science Resources*, books, FOSSweb interactives, Internet, interviews, videos, experiments, etc.).
5. Reevaluate the problem or question based on what has been learned.
6. Revise the idea or conjecture.
7. Make presentations (reports, posters, electronic presentations, etc.).
8. Identify new needs, and make new plans.

This process can continue for as long as new ideas and questions occur, or students can present a final product in any of the suggested formats.

Table 1. Examples of how language functions are exercised in science notebook writing and examples of sentence frames and language structures students may use

Language function	Language objectives for writing in notebooks	Language forms, structures, and scaffolds for writing
Organizational setup		
Organize	Set up and organize notebook: table of contents, glossary or index, page numbers, date, turning to the next blank page.	Use structures such as rows, columns, blocks, numbers, location, alphabetizing.
Notebook component—planning the investigation		
Design Strategize	Write a narrative plan: communicate ideas on an approach to answer the focus question or challenge posed in the investigation.	Use adverbs such as *first, second, next, then, finally*. Brainstorming ideas: First I will ____, and then I will ____. I will need to ____ to ____.
List Record	Write ordered lists: materials, variables, vocabulary words, bullets	We need ____, ____, ____, and ____ to ____ .
Sequence	Record step-by-step procedures: explain exactly what to do, number steps in a sequence, specifics for measurement, how controls will be established, how variables will be measured (tools/ units).	1. ____________. 2. ____________. 3. ____________. I will change ____. I will not change ____. I will measure ____. I will not measure ____.
Notebook component—data acquisition		
Describe	Write narratives: use details, sensory observations, connections to prior knowledge.	I observed ____. When I touch the ____, I feel ____. The ____ has ____. I noticed ____. It feels ____. It smells ____. It sounds ____. It reminds me of ____, because ____.

Table 1 (*continued*)

Language function	Language objectives for writing in notebooks	Language forms, structures, and scaffolds for writing
Notebook component—data acquisition (continued)		
Draw Label Identify	Make technical drawings: draw large, accurate, and detailed representations; identify parts of a system.	Label drawing, using science vocabulary. Recognize shapes, form, location, color, size, and scale. My drawing shows ____.
Organize Compare Classify	Make charts and tables: use a T-table or chart for recording and displaying data.	Set up rows, columns, headings. My T-table compares ____.
Sequence Compare	Record changes: use language structures to communicate change over time, cause and effect.	At first, ____, but now ____. We saw that first ____, then ____, and finally ____. When I ____, it ____. After I ____, it ____.
Notebook component—data organization		
Enumerate	Plot graphs: decide when and how to use bar graphs, line plots, and two-coordinate graphs to organize data.	Define x- and y-axes; provide axis labels, title, units, coordinates; use equal intervals, set origin = (0, 0).
Compare Classify Sequence	Use graphic organizers and narratives to express similarities and differences, to assign an object or action to the category or type to which it belongs, and to show sequencing and order.	This ____ is the same as ____ because ____. This ____ is different than ____ because ____. All these are ____ because ____. ____, ____, and ____ all have/are ____.
Analyze	Use graphic organizers, narratives or concept maps to identify part/whole or cause-and-effect relationships.	Use relationship verbs such as *contain*, *consist of*. As ____, then ____. When I changed ____, then ____ happened. The more/less ____, then ____.

Table 1 (*continued*)

Language function	Language objectives for writing in notebooks	Language forms, structures, and scaffolds for writing
Notebook component—making sense of data		
Infer Explain	Provide claims and evidence: write assertions about what was learned from the investigation, use the data as evidence to support those claims.	Use inferential logical connectors such as *although, while, thus, therefore*. I claim that ____. I know this because ____.
Provide evidence	Use qualitative and quantitative data from the investigation as evidence to support claims.	Use qualitative descriptors such as *more/less, longer/shorter, brighter/dimmer*. Use quantitative expressions using standard metric units of measurement such as cm, mL, °C. My data show ____.
Summarize Predict Generalize	Write a summary narrative to communicate what was learned; ask questions and make predictions based on the newly acquired knowledge.	Answer the focus question by rewriting it as a statement and providing evidence from data. Make a concluding statement. I learned ____. Therefore, I think ____. I predict ____ because ____. A new question I have is ____.
Notebook component—next-step strategies		
Critique Evaluate	Reflect on experience: review notebook entries and revise, using line of learning or 3 Cs (correct in red, confirm in green, and complete in blue).	I used to think ____, but now I think ____. I have changed my thinking about ____. I am confused about ____ because ____. I wonder ____.

READING DOMAIN

Reading is an integral part of science learning. Just as scientists spend a significant amount of their time reading one another's published works, students need to learn to read scientific text—to read effectively for understanding with a critical focus on the ideas being presented.

The articles in *FOSS Science Resources* help facilitate sense making as students make connections to the science concepts introduced and explored during the active investigations. Concept development is most effective when students are allowed to experience organisms, objects, and phenomena firsthand before engaging the concepts in text. The text and illustrations help students make connections between what they have experienced concretely and the abstract ideas that explain their observations.

FOSS Science Resources supports developing literacy skills by providing reading material that corresponds exactly to the concrete, personal experience provided in the active investigations. Students read with enthusiasm when they recognize familiar materials, organisms, and activities and are eager to tackle the reading to confirm their prior knowledge and discover more about the topic. In addition to making connections, once engaged, students naturally use other reading-comprehension strategies, such as asking questions, visualizing, inferring, and synthesizing, to help them understand the reading. As students apply these strategies, they are, in effect, using some of the same scientific thinking processes that promote critical thinking and problem solving.

Reading in the Primary Grades

In the kindergarten modules, you can enhance science learning by using trade books and other read-aloud resources to engage students and provide topics for lively discussions. Reading aloud helps primary students understand the science content and lets you model reading-comprehension strategies, such as asking yourself questions (thinking aloud) and summarizing a paragraph just read. The Reading in *FOSS Science Resources* sections (in every part that has a reading) usually offer suggestions for activating prior knowledge before reading, indicate places to pause and discuss key points during the reading, and describe activities to deepen understanding after the reading. As students develop their reading skills, you might try these different ways to read from *FOSS Science Resources*.

- Read aloud from the big book while students follow along in their own books.
- Lead students in small guided reading groups.
- Have students read aloud with a partner.
- Have students read silently on their own.

The same strategies used in language arts can be applied to reading in science. Begin with reading the article aloud so that students can hear the content read fluently and listen for meaning and coherence. Go back and review the questions within the article. Use think-pair-share or other discussion protocols to allow students to think first, share with a partner, and then respond to the group. Emphasize blending phonemes as you read the text again, and model tracking, connecting spoken words with written words, and helping students identify sight words. Model reading-comprehension strategies by using think-alouds (as you think aloud, you explain the process that you are using in order to understand the text while reading). The expository text structure also provides the opportunity for primary students to learn how to extract information from a table of contents, a glossary, an index, and other text conventions such as headings, subheads, boldface and italic print, labeled graphics, and captions.

At the end of articles, use the questions provided to guide understanding and to assess comprehension and vocabulary acquisition. Choose one or two questions for students to answer in their notebooks. Emphasize the importance of science vocabulary and the appropriate language forms and structures. Here are some ways students can enhance reading comprehension.

- Have students predict the sequence of events or content.
- Have students write or dictate questions about the text and illustrations.
- Use visualization with students to "see, touch, feel, smell, hear" in their minds the content presented in the article.
- Ask students to make connections to information in the article and their observations during the active investigation.

Reading in the Upper-Elementary Grades

As students progress from *learning to read* to *reading to learn*, they apply the strategies and skills of reading comprehension to learning about science in *FOSS Science Resources* and other texts. Writing becomes increasingly important in helping students make sense of the readings as well as the active investigations. Use the suggested questions in the *Investigations Guide* to support comprehension as students read from *FOSS Science Resources*. For most of the investigation parts, the articles are designed to follow the active investigation and are interspersed throughout the flow of the module. This allows students to acquire the necessary background knowledge through active experience before tackling the wider-ranging content and relationships presented in the text. Additional strategies for reading are derived from the seven essential strategies that readers use to help them understand what they read (Keene and Zimmermann 2007).

- Monitor for meaning: Discover when you know and when you don't know.
- Use and create schemata: Make connections between the novel and the known; activate and apply background knowledge.
- Ask questions: Generate questions before, during, and after reading that reach for deeper engagement with the text.
- Determine importance: Decide what matters most, what is worth remembering.
- Infer: Combine background knowledge with information from the text to predict, conclude, make judgments, and interpret.
- Use sensory and emotional images: Create mental images to deepen and stretch meaning.
- Synthesize: Create an evolution of meaning by combining understanding with knowledge from other texts/sources.

Following are some strategies that enhance the reading of expository texts in general and have proven to be particularly helpful in science.

Build on background knowledge. Activating prior knowledge is critical for helping students make connections between what they already know and new information. Reading comprehension improves when students have the opportunity to think, discuss, and write about what they know about a topic before reading. Review what students learned from the active investigation, provide prompts for making connections, and ask questions to help students recall past experiences and previous exposure to concepts related to the reading.

Strategies for Reading in Upper-Elementary Grades

- *Build on background knowledge*
- *Create an anticipation guide*
- *Draw attention to vocabulary*
- *Preview the text*
- *Turn and talk*
- *Jigsaw text reading*
- *Note making*
- *Interactive reading aloud*
- *Summarize and synthesize*
- *3-2-1*
- *Write reflections*
- *Preview and predict*
- *SQ3R*

Create an anticipation guide. Create true-or-false statements related to the key ideas in the reading selection. Ask students to indicate if they agree or disagree with each statement before reading, then have them read the text, looking for the information that supports their true-or-false claims. Anticipation guides connect students to prior knowledge, engage them with the topic, and encourage them to explore their own thinking.

Draw attention to vocabulary. Check the article for bold words to determine if there are words students may not know. Review the science words that have already been listed on the word wall and in students' notebooks. For new science and nonscience vocabulary words that appear in the reading, have students predict their meanings before reading. During the reading, have students use strategies such as context clues and word structure to see if their predictions were correct. This strategy activates prior knowledge and engages students by encouraging analytical participation with the text.

NOTE
Discourse markers are words or phrases that relate one idea to another. Examples are "however," "on the other hand," and "second."

Preview the text. Give students time to skim through the selection, noting subheads, before reading thoroughly. Point out the particular structure of the text and what discourse markers to look for. For example, most *FOSS Science Resources* articles are written as cause and effect, problem and solution, question and answer, comparison and contrast, description, and sequence. Students will have an easier time making sense of the text if they know what to look for in terms of text structure. Model and have students practice analyzing these different types of expository text structures by looking for examples, patterns, and the discourse markers.

Point out how *FOSS Science Resources* text is organized (titles, headings, subheadings, questions, and summaries) and how to use the table of contents, glossary, and index. Explain how to scan for formatting features that provide key information (such as bold type and italics, captions, and framed text) and graphic features (such as tables, graphs, photographs, maps, diagrams, charts, and illustrations) that help clarify, elaborate, and explain important information in the reading.

While students preview the article, have them focus on the questions that appear in the text, as well as questions at the end of the article. Encourage students to write down questions they have that they think the reading will answer.

Turn and talk. When reading as a whole class, stop at key points and have students share their thinking about the selection with the student sitting next to them or in their collaborative group. This strategy helps students process the information and allows everyone to participate in the discussion. When reading in pairs, encourage students to stop and discuss with their partners. One way to encourage engagement and understanding during paired reading is to have students take turns reading aloud a paragraph or section on a certain topic. The one who is listening then summarizes the meaning conveyed in the passage.

Jigsaw text reading. Students work together in small groups (expert teams) to develop a collective understanding of a text. Each expert team is responsible for one portion of the assigned text. The teams read and discuss their portions to gain a solid understanding of the key concepts. They might use graphic organizers to refine and organize the information. Each expert team then presents its piece to the rest of the class. Or new, small jigsaw groups can be formed that consist of at least one representative from each expert team. Each student shares with the jigsaw group what their team learned from their particular portion of the text. Together, the participants in the jigsaw group fit their individual pieces together to create a complete picture of the content in the article.

Note making. The more interactive that students make a reading, the better for their understanding. Encourage students to become active readers by asking them to make notes as they read. Studies have shown that note making—especially paraphrasing and summarizing—is one of the most effective means for understanding text (Graham and Herbert 2010; Applebee 1984).

Students can annotate the text by writing thoughts and questions on self-stick notes. Using symbols or codes can help facilitate comprehension monitoring. Here are some possible symbols (Harvey 1998).

- ★ interesting
- BK background knowledge
- ? question
- C confusing
- I important
- L learning something new
- W wondering
- S surprising

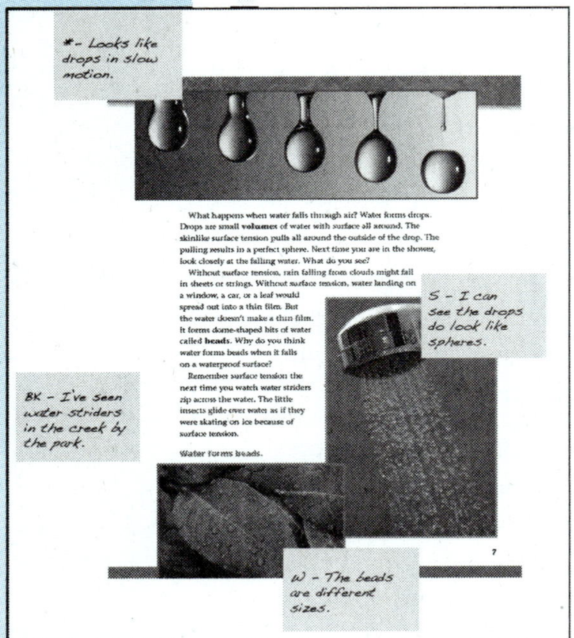

What happens when water falls through air? Water forms drops. Drops are small **volumes** of water with surface all around. The skinlike surface tension pulls all around the outside of the drop. The pulling results in a perfect sphere. Next time you are in the shower, look closely at the falling water. What do you see?

Without surface tension, rain falling from clouds might fall in sheets or strings. Without surface tension, water landing on a window, a car, or a leaf would spread out into a thin film. But the water doesn't make a thin film. It forms dome-shaped bits of water called **beads**. Why do you think water forms beads when it falls on a waterproof surface?

Remember surface tension the next time you watch water striders zip across the water. The little insects glide over water as if they were skating on ice because of surface tension.

Water forms beads.

7

FACTS	QUESTIONS
C - Surface tension makes water form into drops.	Why does a drop always have the shape of a sphere?
I - When the drops land they make a dome shape, called beads.	Why don't I have beads of water on my clothes when it rains?
L - Water Striders skate over the water.	Can water striders glide over other liquids?

An example of annotated text in *FOSS Science Resources*

Students can use a set of symbols while making notes about connections. The readings in *FOSS Science Resources* incorporate the active learning that students gain from the investigations so that they can make authentic text-to-self (T-S) connections. In other words, what they read reminds them of firsthand experiences, making the article more engaging and easier to understand. Text-to-text (T-T) connections are notes students make when they discover a new idea that reminds them of something they've read previously in another text. Text-to-world (T-W) connections involve the text and more global everyday connections to students' lives.

You can model note-making strategies by displaying a selection of text, using a projection system, a document camera, or an interactive whiteboard. As you read the text aloud, model how to write comments on self-stick notes, and use a graphic organizer in a notebook to enhance understanding.

Graphic organizers help students focus on extracting the important information from the reading and analyzing relationships between concepts. This can be done by simply having students make columns in their notebooks to record information and their thinking (Harvey and Goudvis 2007). Here are two examples of graphic organizers.

Notes	Thinking

Facts	Questions	Responses

Interactive reading aloud. Reading aloud is an effective strategy for enhancing text comprehension. It offers opportunities to model specific reading-comprehension strategies and allows students to concentrate on making sense of the content. When modeling, share the thinking processes used to understand the reading (questioning, visualizing, comparing, inferring, summarizing, etc.), then have students share what they observed you thinking about as an active reader.

Summarize and synthesize. Model how to pick out the important parts of the reading selection. Paraphrasing is one way to summarize. Have students record summaries of the reading, using their own words. To scaffold the learning, use graphic organizers to compare and contrast, group, sequence, and show cause and effect. Another method is to have students make two columns in their notebooks. In one column, they record what is important, and in the other they record their personal responses (what the reading makes them think about). When writing summaries, tell students,

- *Pick out the important ideas.*
- *Restate the main ideas in your own words.*
- *Keep it brief.*

3-2-1. This strategy gives students the opportunity to synthesize information and formulate questions they still have regarding the concepts covered in an article. In their notebooks, students write three new things they learned, two interesting things worth remembering and sharing, and one question that occurred to them while reading the article. Other options might include three facts, two interesting ideas, and one insight about themselves as learners; three key words, two new ideas, and one thing to think about (modified from Black Hills Special Service Cooperative 2006).

Write reflections. After reading, ask students to review their notes in their notebooks to make any additions, revisions, or corrections to what they recorded during the reading. This review can be facilitated by using a line of learning. Students draw a line under their original conclusion or under their answer to a question posed at the end of an article. They then add any new information as a new narrative entry. The line of learning indicates that what follows represents a change of thinking.

Preview and predict. Instruct students to independently preview the article, directing attention to the illustrations, photos, boldfaced words, captions, and anything else that draws their attention. Working with a partner, students discuss and write three things they think they will learn from the article. Have partners verbally share their list with another pair of students. The group of four can collaborate to generate one list. Teams report their ideas, and together you create a class list on chart paper.

Read the article aloud, or have students read with a partner aloud or silently. Referring to the preview/prediction list, discuss what students learned. Have them record the most important thing they learned from the reading for comparison with the predictions.

SQ3R. Survey! Question! Read! Recite! Review! provides an overall structure for before, during, and after reading. Students begin by previewing the text, looking for features that will help them make predictions about content. Based on their surveys, students develop questions to answer as they read. Students then read the selection. They recite by telling a partner what they've learned (their partner listens carefully to their recitation, making sure they haven't missed any important concepts) and answering any questions their partner might have. Finally, they review their questions and answers. There is value in asking students to make their entire SQ3R process explicit.

Use this script as an outline, using appropriate wording where necessary.

Survey! Question!

a. Before you read, survey (preview) the reading selection.
b. As you survey, think of questions that the reading might answer while you read the title, headings, and subheadings.
c. Read the questions at the end of the article.
d. Ask yourself, "What did my teacher say about this selection when it was assigned?"
e. Ask yourself, "What do I already know about this subject from experiments or discussions we've had in class?"

Read!

a. While reading, look for answers to the questions you raised.
b. Think about answers for the questions at the end of the article.
c. Study graphics, such as pictures, graphs, and tables.
d. Reread captions associated with pictures, graphs, and tables.
e. Note all italicized and boldfaced words or phrases.
f. Reduce your reading speed for difficult passages.
g. Stop and reread parts that are not clear.
h. Read only a section at a time, and recite after each section.

Recite!

a. After you've read a section, make notes in your science notebook by writing information in your own words.
b. Underline or highlight important notes in your science notebook.
c. Use the method of recitation that best suits your particular learning style. Remember that the more senses you use, the more likely you are to remember what you read. Seeing, saying, hearing, and writing will all enhance learning.
d. Ask questions aloud about what you just read, and summarize aloud, in your own words, what you read.
e. Listen attentively as your partner recites.

Review!

Reviewing should be ongoing. With the whole class, review the entire process, calling on groups to talk to the class about what they've learned.

Struggling Readers

For students reading below grade level, the strategies listed above can be modified to support reading comprehension by integrating strategies from the Reading in the Primary Grades section, such as read-alouds and guided reading. Students can also follow along with the audio stories on FOSSweb. Breaking the reading down into smaller chunks, providing graphic organizers, and modeling reading-comprehension strategies can also help students who may be struggling with the text. For additional strategies for English learners, see the supported-reading strategy in the English-Language Development section of this chapter.

SCIENCE-VOCABULARY DEVELOPMENT

Words play two critically important functions in science. First and most important, we play with ideas in our minds, using words. We present ourselves with propositions—possibilities, questions, potential relationships, implications for action, and so on. The process of sorting out these thoughts involves a lot of internal conversation, internal argument, weighing of options, and complex linguistic decisions. Once our minds are made up, communicating that decision, conclusion, or explanation in writing or through verbal discourse requires the same command of the vocabulary. Words represent intelligence; acquiring the precise vocabulary and the associated meanings is key to successful scientific thinking and communication.

The words introduced in FOSS investigations represent or relate to fundamental science concepts and should be taught in the context of the investigation. Many of the terms are abstract and are critical to developing science content knowledge and scientific and engineering practices. The goal is for students to use science vocabulary in ways that demonstrate understanding of the concepts the words represent—not to merely recite scripted definitions. The most effective science-vocabulary development strategies help students make connections to what they already know. These strategies focus on giving new words conceptual meaning through experience; distinguishing between informal, everyday language and academic language; and using the words in meaningful contexts.

Building Conceptual Meaning through Experience

In most instances, students should be presented with new words in the context of the active experience at the need-to-know point in the investigation. Words such as *flexible*, *magnetic*, *viscous*, *silt*, *erode*, and *electricity* are conceptually loaded and significantly abstract. Students will have a much better chance of understanding, assimilating, and remembering the new word (or new meaning) if they can connect it with a concrete experience.

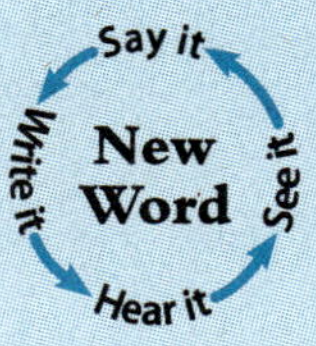

The new-word icon appears in the sidebar when you introduce a word that is critical to understanding the concepts or scientific practices students will be learning and applying in the investigation. When you introduce a new word, students should

- Hear it: Students listen as you model the correct contextual use and pronunciation of the word.
- See it: Students see the new word written out. Add a visual reference (an illustration or a sample) next to the word if possible.
- Say it: Have primary-grade students repeat the word chorally and clap out the syllables.
- Write it: You write the word on the board, chart paper, sentence strip, or card. Students use the new words in context when they write in their notebooks.
- Act it: Demonstrate action words such as *separate*, *compare*, and *observe* (using total physical response).

Bridging Informal Language to Science Vocabulary

Students bring a wealth of language experience to the classroom. FOSS investigations are designed to tap into students' inquisitive natures and their excitement of discovery in order to encourage lively discussions as they explore materials in creative ways. There should be a lot of talking during science time! Your role is to help students connect informal language to the vocabulary used to express specific science concepts. As you circulate during active investigation, you continually model the use of science vocabulary. For example, when a student holds up a bottle of water and says, "I can see through it!" you might respond, "Yes, I see what you mean; the liquid in the bottle is transparent." Following are some strategies for validating students' conversational language while developing their familiarity with and appreciation for science vocabulary.

Word bubbles. Choose a word from the word wall that is widely used by students and that is a synonym for a science vocabulary word. Draw a circle on the board or chart paper, and write the word in the center. Draw lines out from the circled word, and make more circles. Ask students to call out more synonyms for the word, and write them in the outer circles. Introduce the target vocabulary word as yet one more synonym for the target word, a word that is used in science. Highlight the word, and model its correct usage and pronunciation. Encourage students to use it in their discussions and in their notebook entries. Introduce the science word that is its opposite, when appropriate.

Bridging Language Strategies

- *Word bubbles*
- *Word sorts*
- *Semantic webs*
- *Concept maps*
- *Cognitive content dictionaries*
- *Word associations*

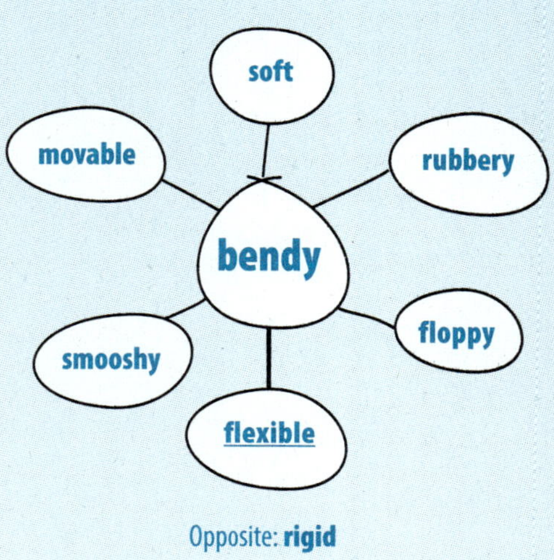

Word bubbles

Word sorts. Make a set of word cards from words on the word wall. Ask students to help you group the words that are synonyms or that have conceptual connections. Add the new science words to the card set. Repeat the process with a new set of words. For upper-elementary students, make sets of vocabulary cards for students to sort in small groups.

Semantic webs. Select a vocabulary word, and write it in the center of a piece of paper (or on the board if doing this with the whole class). Brainstorm a list of words or ideas that are related to the first word. Group the words and concepts into several categories, and attach them to the central word with lines, forming a web (modified from Hamilton 2002).

Concept maps. Select six to ten related science words. Have students write them on self-stick notes or cards. Have small groups discuss how the words/concepts are related. Students organize words in groups and glue them down or copy them onto a sheet of paper. Students draw lines between the related words. On the lines, they write words describing or explaining how the concept words are related.

Cognitive content dictionaries. There are many variations of frameworks to help students record new words. This example can be used with the whole class with primary students or individually with upper-elementary students to introduce a few key vocabulary words used in an investigation. Have students write the word, predict its meaning, write the final meaning after class discussion (using primary language or an illustration when appropriate), and then use the word in a sentence. The word can then be used as a signal word to call for attention.

Cognitive Content Dictionary	
New word	esker
Prediction (clues)	a flower growing at the North Pole
Final meaning (e.g., primary language, pictures)	a sand and gravel ridge, formed at the edges of a stream, flowing under a glacier
How I would use it (sentence)	The esker showed that a glacier had been in the valley at some time in the past.

Word associations. In this brainstorming activity, you say a word, and students respond by writing the first word that comes to mind. Then students share their words with the class. This activity builds connections to students' prior frames of reference.

Using Science Vocabulary in Context

In order for a new vocabulary word to become part of a student's functional vocabulary, he or she must have ample opportunities to hear and use it. The use of vocabulary terms is embedded in the activities through teacher talk, whole-class and small-group discussions during investigations, writing in science notebooks, readings, assessments, and games. In addition, other methods used during language-arts instruction can be used to reinforce important vocabulary words and phrases.

Science Vocabulary Strategies

- *Word wall/word cards*
- *Drawings and diagrams*
- *Cloze activities*
- *Word wizard*
- *Word analysis/word parts*
- *Breaking apart words*
- *Possible sentences*
- *Reading*
- *Highlighting the vocabulary*
- *Index*
- *Poems, chants, and songs*
- *Games*

Word wall/word cards. Use chart paper or a pocket chart to record both science content and procedural words. Record the words as they come up during and after the investigations. Then copy key words on sentence strips or cards, and put them in a pocket chart. With a pocket chart, words can be sorted and moved around easily. For example, you could ask students to find words that are synonyms, antonyms, nouns, or verbs. Word cards should be available to each group during the investigation. This allows students to retrieve a word quickly when they are labeling diagrams and objects used during the investigation.

Drawings and diagrams. For English learners and visual learners, a diagram can be used to review and explain abstract content. Ahead of time, draw an illustration lightly, almost invisibly, with pencil on chart paper. When it's time for the investigation, trace the illustration with markers as you introduce the words and phrases to students. Students will be amazed by your artistic ability.

Cloze activities. Structure a sentence for students to complete, leaving out the vocabulary word, and crafting the sentence so that the missing vocabulary word is the last word in the sentence. You can do this chorally with primary students or in writing on the board or chart paper for upper-elementary students. Here's an example from the **Solids and Liquids Module**.

> Teacher: *Liquids that are clear and that you can see through are _____.*
>
> Students: *Transparent.*

Word wizard. Tell students that you are going to lead a word activity. You will be thinking of a science vocabulary word from the word wall. The goal is to figure out the word. Provide hints that have to do with parts of a definition, root word, prefix, suffix, and other relevant components. Students work in teams of two to four. Provide one hint, and give teams 1 minute to discuss. One team member writes the word on a piece of paper or on the whiteboard, using dark marking pens. Each team holds up its word for only you to see. After the third clue, reveal the word, and move on to the next word.

1. *This word is part of a plant.*
2. *It is usually not green.*
3. *It brings water and nutrients into the plant.*

It is the **root**.

Word analysis/word parts. Learning clusters of words that share a common origin can help students understand content-area texts and connect new words to familiar ones. This type of contextualized teaching meets the immediate need of understanding an unknown word while building generative knowledge that supports students in figuring out difficult words for future reading.

geology

geologist

geological

geography

geometry

geophysical

Breaking apart words. Have teams of two to four students break the word into prefix, root word, and suffix. Give each team different words, and have each team share the parsed elements of the word with the whole class.

electromagnetism

electro: having to do with electricity

magnet: having polar properties; attraction and repulsion of opposite and similar poles

ism: relating to a theory of how things behave

Possible sentences. Here is a simple strategy for teaching word meanings and generating class discussion.

1. Choose six to eight key concept words from the text of an article in *FOSS Science Resources*.
2. Choose four to six additional words that students are more likely to know something about.
3. Put the list of 10–14 words on the board or project it. Provide brief definitions as needed.
4. Ask students to devise sentences that include two or more words from the list.
5. On chart paper, write all sentences that students generate, both coherent and otherwise.
6. Have students read the article from which the words were extracted.
7. Revisit students' sentences, and discuss whether the sentences are sensible based on the passage or how they could be modified to be more coherent.

Reading. After the active investigation, students continue to develop their understanding of the vocabulary words and the concepts those words represent by listening to you read aloud, reading with a partner, or reading independently. Use strategies discussed in the Reading Domain section to encourage students to articulate their thoughts and practice the new vocabulary.

Highlighting the vocabulary. Emphasize the vocabulary words students should be using when they answer the focus question in their science notebooks. Distribute copies of the vocabulary/glossary for the investigation (available on FOSSweb) for students to glue into their notebooks. As you introduce the words in the investigation, students highlight them.

Index. Have students create an index at the back of their notebooks. There, they can record new vocabulary words and the notebook page where they defined and used the new words for the first time in the context of the investigation.

NOTE
See the Science Notebooks in Grades 3–5 chapter for an example of this index.

Poems, chants, and songs. Vocabulary words and phrases can be reinforced using content-rich poems, rhymes, chants, and songs.

Games. The informal activities included in the investigations are designed to reinforce important vocabulary words. Once students learn them, the words can be integrated into any type of independent work time, such as centers, workshops, and early-finisher tasks.

NOTE
English-language development refers to the advancement of students' ability to read, write, and speak English.

EL NOTE
Look for EL Notes in the investigations at points where students at beginning levels of English proficiency may need additional support.

ENGLISH-LANGUAGE DEVELOPMENT

Active investigations, together with ample opportunities to develop and use language, provide an optimal learning environment for English learners. This section highlights the English-language development (ELD) opportunities inherent in FOSS investigations and suggests other best practices for facilitating both the learning of new science concepts and the development of academic vocabulary and language structures that enhance literacy. For example, the hands-on structure of FOSS investigations is essential for the conceptual development of science content knowledge and the habits of mind that guide and define scientific practices. Students are engaged in concrete experiences that are meaningful and that provide a shared context for developing understanding—critical components for ELD instruction.

To further address the needs of English learners, the *Investigations Guide* includes EL Notes at points in the investigations where students at beginning levels of English proficiency may need additional support. When getting ready for an investigation, review the EL Notes, and determine the points where English learners may require scaffolds and where the whole class might benefit from additional language-development supports. One way to plan for ELD integration in science is to keep in mind four key areas: prior knowledge, comprehensible input, academic language development, and oral practice. The ELD chart below lists examples of universal strategies for each of these components that work particularly well in teaching science.

English-Language Development (ELD) Quadrants	
Activating prior knowledge • Inquiry chart • Circle map • Observation poster • Quick write • Kit inventory	**Using comprehensible input** • Content objectives • Multiple exposures • Visual aids • Supported reading • Procedural vocabulary
Developing academic language • Language objectives • Sentence frames • Word wall, word cards, drawings • Concept maps • Cognitive content dictionaries	**Providing oral practice** • Small-group discussions • Science talk • Oral presentations • Poems, chants, and songs • Teacher feedback

Activating Prior Knowledge

When an investigation engages a new concept, first students recall and discuss familiar situations, objects, or experiences that relate to and establish a foundation for building new knowledge and conceptual understanding. Eliciting prior knowledge also supports learning by motivating interest, acknowledging culture and values, and checking for misconceptions and prerequisite knowledge. This is usually done in the first steps of Guiding the Investigation in the form of an oral discussion, presentation of new materials, or a written response to a prompt. The tools outlined below can also be used before beginning an investigation to establish a familiar context for launching into new material.

Circle maps. Draw two concentric circles on chart paper. In the middle circle, write the topic to be explored. In the second circle, record what students already know about the subject. Ask students to think about how they know or learned what they already know about the topic. Record the responses outside the circles. Students can also do this independently in their science notebooks.

An example of a circle map

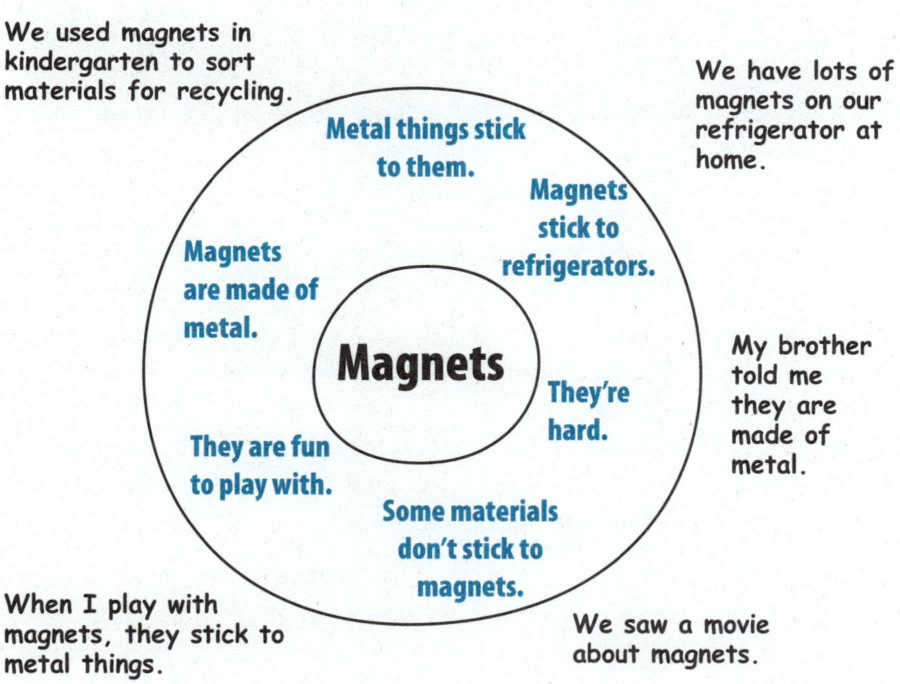

Strategies for Activating Prior Knowledge

- *Circle maps*
- *Observation posters*
- *Quick writes*
- *Kit inventories*

Science-Centered Language Development

TEACHING NOTE

The images selected for the observation posters should pique students' interest but not take away from discoveries that students will make when working with the materials.

Observation posters. Make observation posters by gluing or taping pictures and artifacts relevant to the module or a particular investigation onto pieces of blank chart paper or poster paper. Hang them on the walls in the classroom, and have students rotate in small groups to each poster. At each station, students discuss their observations with their partners or small groups and then record (write, draw, or dictate) an observation, a question, a prediction, or an inference about the pictures as a contribution to the commentary on the poster.

An observation poster

Quick writes. Ask students what they think they know about the topic of the investigation. Responses can be recorded independently as a quick write in science notebooks and then shared collaboratively. You should not correct misconceptions initially. Periodically revisit the quick-write ideas as a whole class, or have students review their notebook entries to correct, confirm, or complete their original thoughts as new information is acquired (possibly using a line of learning). At the conclusion of the investigation, students should be able to express mastery of the new conceptual material.

Kit inventories. Introduce each item from the FOSS kit used in the investigation, and ask students questions to get them thinking about what each item is and where they may have seen it before. Have them describe the objects and make predictions about how they will be used in the investigation. Tape samples of the items on chart paper, or print and display the equipment photo cards (download from FOSSweb) along with the name and a description, to serve as an interactive word wall.

Item	Teacher	Student
Bottle 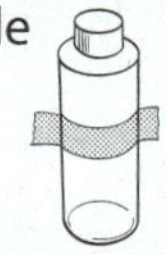	What is this? What is it made of? What is it used for?	A bottle. Plastic. To hold liquids.
Scoop 	What is this? It's like a spoon. It's called a scoop. What do you think we will be using it for in science?	It looks like a spoon. Like an ice-cream scoop! To scoop things up.
Funnel 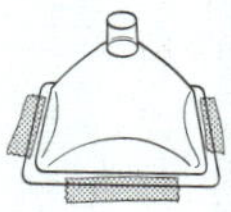	Have you seen this before? It's called a funnel. Where else have you seen this? Can you describe it?	My mom uses that for the car. In the kitchen. My uncle uses it sometimes to pour things. It's round on the ends. It's bigger on one end, and it's hollow.
Beaker 	Have you seen this before? It's called a beaker. Where else have you seen this?	We used it for science last year to measure and pour water.
Vial 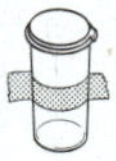	This is a vial. What do you think we will be using it for in science?	To hold small things.

A kit inventory script from the Solids and Liquids Module

Comprehensible Input

In order to initiate their own sense-making process, students must be able to access the information presented to them. We refer to this ability as comprehensible input. Students must understand the essence of new ideas and concepts before beginning the process of constructing new scientific meaning. The strategies for comprehensible input used in FOSS ensure that the delivery of instruction is understandable while providing students with the opportunity to grapple with new ideas and the critically important relationships between concepts. Additional tools such as repetition, visual aids, emphasis on procedural vocabulary, and auditory reinforcement can also be used to enhance comprehensible input for English learners.

Strategies for Comprehensible Input

- *Content objectives*
- *Multiple exposures*
- *Visual aids*
- *Supported reading*
- *Procedural vocabulary*

Content objectives. The focus question for each investigation part frames the activity objectives—what students should know or be able to do at the end of the part. Making the learning objectives clear and explicit helps English learners prepare to process the delivery of new information, and helps you maintain the focus of the investigation. Write the focus question on the board, have students read it aloud and transcribe it into their science notebooks, and have students answer the focus question at the end of the investigation part. You then check their responses for understanding.

Procedural Vocabulary
Add
Analyze
Assemble
Attach
Calculate
Change
Classify
Collect
Communicate
Compare
Connect
Construct
Contrast
Describe
Demonstrate
Determine
Draw
Evaluate
Examine
Explain
Explore
Fill
Graph
Identify
Illustrate
Immerse
Investigate
Label
List
Measure
Mix
Observe
Open
Order
Organize
Pour
Predict
Prepare
Record
Represent
Scratch
Separate
Sort
Stir
Subtract
Summarize
Test
Weigh

Multiple exposures. Repeat the activity as a class, at a center during independent work time, or in an analogous but slightly different context, ideally one that incorporates elements that are culturally relevant to students.

Visual aids. On the board or chart paper, write out the steps for conducting the investigation. This will provide a visual reference. Include illustrations if necessary. Use graphic representations (illustrations drawn and labeled in front of students) to review the concepts explored in the active investigations. In addition to the concrete objects included in the kit, use realia to augment the activity to help English learners build understanding and make cultural connections. Graphic organizers (webs, Venn diagrams, T-tables, flowcharts, etc.) aid comprehension by helping students see how ideas (concepts) are related.

Supported reading. In addition to the reading comprehension strategies suggested in the Reading Domain section of this chapter, English learners can also benefit from methods such as front-loading key words, phrases, and complex text structures before reading; using preview-review (main ideas are previewed in the primary language, read in English, and reviewed in the primary language); and having students use sentence frames specifically tailored to record key information and/or graphic organizers that make the content and the relationship between concepts visually explicit from the text as they read.

Procedural vocabulary. Make sure students understand the meaning of the words used in the directions describing what they should be doing during the investigation. These may or may not be science-specific words. Use techniques such as modeling, demonstrating, and body language (gestures) to explain procedural meaning in the context of the investigation. The words students will encounter in FOSS include those listed in the sidebar. To build academic literacy, English learners need to learn the multiple meanings of these words and their specific meanings in the context of science.

Developing Academic Language

As students learn the nuances of the English language, it is critical that they build proficiency in academic language in order to participate fully in the cognitive demands of school. Academic language refers to the more abstract, complex, and specific aspects of language, such as the words, grammatical structure, and discourse markers that are needed for higher cognitive learning. FOSS investigations introduce and provide opportunities for students to practice using the academic vocabulary needed to access and meaningfully engage with science ideas.

Language objectives. Consider the language needs of English learners and incorporate specific language-development objectives that will support learning the science content of the investigation, such as a specific word knowledge skill (a way to expand use of vocabulary by looking at root words, prefixes, and suffixes), a linguistic pattern or structure for oral discussion and writing, or a reading-comprehension strategy. Recording in students' science notebooks is a productive place to optimize science learning and language objectives. For example, in the **Pebbles, Sand, and Silt Module**, one language objective might be "Students will describe what happened when they rubbed their rocks together (think-pair-share) and answer the focus question in their notebooks, using a cause–and–effect sentence frame."

Vocabulary development. The Science-Vocabulary Development section in this chapter describes the ways science vocabulary is introduced and developed in the context of an active investigation and suggests methods and strategies that can be used to support vocabulary development during instruction in English language arts and ELD. In addition to science vocabulary, students also need to learn the nonspecific-content words that facilitate deeper understanding and communication skills. Words such as *release*, *convert*, *beneficial*, *produce*, *receive*, *source*, and *reflect* are words used in the investigations and *FOSS Science Resources* and are frequently used in other content areas. Learning these academic vocabulary words gives students a more precise and complex way of practicing and communicating productive thinking. Consider using the strategies described in the Science-Vocabulary Development section to explicitly teach targeted, high-leverage words that can be used in multiple ways and that can help students make connections to other words and concepts. Sentence frames, word wall, concept maps, and cognitive content dictionary are strategies that have been found to be effective with academic-vocabulary development.

NOTE
For additional resources and updated references, go to FOSSweb.

REFERENCES

Applebee, A. 1984. "Writing and Reasoning." *Review of Educational Research* 54 (winter): 577–596.

Bereiter, C. 2002. *Education and Mind in the Knowledge Age*. Hillsdale, NJ: Erlbaum.

Black Hills Special Services Cooperative. 2006. "3-2-1 Strategy." In *On Target: More Strategies to Guide Learning*. Rapid City, SD. http://www.sdesa6.org/content/resources.htm.

Dyasi, H. M. 2006. "Visions of Inquiry: Science." In *Linking Science and Literacy in the K–8 Classroom*, ed. R. Douglas, K. Worth, and W. Binder. Arlington, VA: NSTA Press.

Gibbons, P. 2002. *Scaffolding Language, Scaffolding Learning*. Portsmouth, NH: Heinemann.

Graham, S., and M. Herbert. 2010. *Writing to Read: Evidence for How Writing Can Improve Reading*. New York: Carnegie.

Hamilton, G. 2002. *Content-Area Reading Strategies: Science*. Portland, ME: Walch Publishing.

Hand, B., and V. Prain. 2002. "Teachers Implementing Writing-to-Learn Strategies in Junior Secondary Science: A Case Study." *Science Education* 86: 737–755.

Harvey, S. 1998. *Nonfiction Matters: Reading, Writing, and Research in Grades 3–8*. Portland, ME: Stenhouse.

Harvey, S., and A. Goudvis. 2007. *Strategies That Work: Teaching Comprehension for Understanding and Engagement*. Portland, ME: Stenhouse.

Keene, E., and S. Zimmermann. 2007. *Mosaic of Thought: The Power of Comprehension Strategies*. 2nd ed. Portsmouth, NH: Heinemann.

Keys, C. 1999. *Revitalizing Instruction in Scientific Genres: Connecting Knowledge Production with Writing to Learn in Science*. Athens: University of Georgia.

Lieberman, G. A., and L.L. Hoody. 1998. *Closing the Achievement Gap: Using the Environment as an Integrating Context for Learning*. San Diego, CA: State Education and Environment Roundtable.

National Research Council. 2012. *A Framework for K–12 Science Education: Practices, Crosscutting Concepts, and Core Ideas*. Committee on Conceptual Framework for New Science Education Standards.

Ostlund, K. 1998. "What the Research Says about Science Process Skills: How Can Teaching Science Process Skills Improve Student Performance in Reading, Language Arts, and Mathematics?" *Electronic Journal of Science Education* 2 (4).

Wellington, J., and J. Osborne. 2001. *Language and Literacy in Science Education*. Buckingham, UK: Open University Press.

Winokur, J., and K. Worth. 2006. "Talk in the Science Classroom: Looking at What Students and Teachers Need to Know and Be Able to Do." In *Linking Science and Literacy in the K–8 Classroom*, ed. R. Douglas, K. Worth, and W. Binder. Arlington, VA: NSTA Press.

FOSS and Common Core ELA – Grade K

FOSS and Common Core ELA — Grade K

Contents

INTRODUCTION

Each FOSS investigation follows a similar design to provide multiple exposures to science concepts. The design includes these pedagogies.

- Active investigation, including outdoor experiences
- Writing in science notebooks to answer focus questions
- Reading in *FOSS Science Resources*
- Assessment to monitor progress and motivate student reflection on learning

In practice, these components are seamlessly integrated into a continuum designed to maximize every student's opportunity to learn. An instructional sequence may move from one pedagogy to another and back again to ensure adequate coverage of a concept.

The FOSS instructional design recognizes the important role of language in science learning. Throughout the pedagogical design elements, students engage in the practices of the Common Core State Standards for English Language Arts. The purpose of this chapter is to provide the big picture of how FOSS provides opportunities for the development and exercising of these practices through science. On the following pages, there is a chart that identifies the opportunities for kindergarten and where the relevant opportunities are found within the three FOSS modules.

Guiding Principles

When integrating language-arts instruction with FOSS, keep in mind these guiding principles:

- FOSS investigations follow a clear and coherent conceptual flow and a consistent instructional design. Students develop science knowledge by building a framework of concepts and supporting ideas.
- Common Core State Standards for ELA are introduced, developed, and practiced in the context of learning science content and engaging in the science and engineering practices. Students read and comprehend complex science texts related to their prior experience and knowledge. They write informational/explanatory texts, arguments to support claims, and narratives about experience in science. They engage in collaborative discussions about science and learn new vocabulary and language structures in context.
- The decision to use additional science texts, writing tasks, oral discourse opportunities, and vocabulary development activities is based on how well they address the science as well as the ELA standards.
- Instruction is differentiated to meet the needs of all students; the linguistic accommodations that are made for English learners support comprehensible input and accelerate academic language development. Language objectives for English learners in science instruction include the application of strategies that support construction of meaning from academic discussions and complex text, participation in productive discourse, and the ability to express ideas in writing clearly and coherently according to task, purpose, and audience.
- Formative assessment tools are used routinely to measure progress toward science understanding, use of science and engineering practices, and meeting literacy and language development goals. Assessment is viewed as a way to make student thinking visible and to determine next steps for instruction for both science and literacy. Instruction includes opportunities for students to assess themselves and peers.

Adhering to these guiding principles optimizes instructional time and, most importantly, benefits student learning by providing authentic and relevant contexts for building content knowledge, applying meaning-making strategies, and developing language and literacy skills.

Kindergarten is a year of wonder and exciting challenges as students expand their learning about the world around them. They engage in interactive experiences with reading and writing, and develop the social norms necessary for productive discourse. They develop their abilities to make meaning from texts read aloud and begin to decode on their own. Their vocabularies increase and they continue to develop their communication skills. Kindergartners are expected to use the science and engineering practices to demonstrate their understanding of the core ideas. To accomplish this, students learn to ask and answer questions about key details in texts, use images to support understanding, describe connections between science ideas, and communicate their thoughts orally and in written form using models, drawings, and emerging writing.

TEACHING NOTE

Throughout the kindergarten FOSS modules, opportunites for addressing the ELA standards have been noted; however, these examples should not be considered the only places for integrating literacy skills.

Instructional Flow

In almost all investigations, the instructional flow is the same and provides these opportunities for effective integration of ELA standards.

- When **setting the context** for the lesson, students activate prior knowledge through class or small-group discussions where they ask and answer questions to get information and clarify understanding (SL 3), or describe familiar places, things, and events (SL 4).
- During the **active investigation**, students are expected to work with partners and in collaborative groups, and to engage in teacher-led discussions where they follow rules for discussion and continue a conversation through mulitiple exchanges (SL 1).
- In the **data management** phase, students make and record observations, and organize data in their notebooks (W 7). The notebook provides a space for students to recall information from experiences and to gather information to answer the focus question (W 8) and to use words and phrases acquired through conversations and readings (L 6).
- The **analysis** phase involves discussing observations, constructing explanations, and engaging in argumentation (SL 1). Students make meaning by drawing, dictating, and writing explanatory texts (W 2), opinion pieces (W 1), or narratives of events (W 3).
- **Reading** articles in *FOSS Science Resources* and other recommended readings provides a plethora of opportunities to address all the kindergarten reading standards for informational text.
- Lastly, the **assessment** tools and next-step strategies for engaging students in high-level critical thinking support the development of the Common Core State Standards capacities of the literate individual: demonstrate independence, build strong content knowledge, comprehend as well as critique, and value evidence.

FOSS and Common Core ELA — Grade K

READING STANDARDS FOR INFORMATIONAL TEXT

	Grade K Standard	Materials and Motion Module
Key Ideas and Details	1. With prompting and support, ask and answer questions about key details in a text.	Discuss articles in *FOSS Science Resources* Inv 1, Part 7, Steps 9, 10 Inv 2, Part 1, Steps 14, 15 Inv 3, Part 2, Steps 10, 11; Inv 3, Part 4, Step 9; Inv 3, Part 5, Steps 1, 2 Inv 4, Part 1, Steps 16, 17 Inv 4, Part 2, Steps 14, 15
Key Ideas and Details	2. With prompting and support, identify the main topic and retell key details of a text.	Discuss and review articles in *FOSS Science Resources* Inv 1, Part 1, Step 20; Inv 1, Part 7, Steps 9, 10 Inv 2, Part 1, Steps 14, 15 Inv 3, Part 2, Step 11; Inv 3, Part 4, Step 9; Inv 3, Part 5, Steps 1-2 Inv 4, Part 1, Steps 16, 17 Inv 4, Part 2, Steps 14, 15
Key Ideas and Details	3. With prompting and support, describe the connection between two individuals, events, ideas, or pieces of information in a text.	Discuss articles in *FOSS Science Resources* Inv 1, Part 1, Step 20; Inv 1, Part 7, Steps 9, 10 Inv 2, Part 1, Steps 14, 15 Inv 3, Part 2, Steps 10, 11; Inv 3, Part 4, Steps 8, 9; Inv 3, Part 5, Step 2 Inv 4, Part 1, Steps 16, 17 Inv 4, Part 2, Steps 14, 15
Craft and Structure	4. With prompting and support, ask and answer questions about unknown words in a text.	All investigations provide opportunities for students to ask and answer questions to determine the meaning of new words and phrases while reading articles in *FOSS Science Resources.* Selected example Inv 2, Part 1, Step 13
Craft and Structure	5. Identify the front cover, back cover, and title page of a book.	Discuss text features in *FOSS Science Resources* Inv 1, Part 1, Step 19

Common Core State Standards for English language arts and literacy in history/social studies science and technical subjects (National Governors Association Center for Best Practices and Council of Chief State School Officers, 2010).

Trees and Weather Module	Animals Two by Two Module
Discuss articles in *FOSS Science Resources* Inv 1, Part 5, Steps 9, 10; Inv 1, Part 6, Step 11 Inv 2, Part 1, Step 1; Inv 2, Part 6, Steps 6, 7 Inv 3, Part 1, Steps 10, 11 ; Inv 3, Part 3, Steps 9, 10, 12 Inv 4, Part 2, Step 8; Inv 4, Part 4, Step 9; Inv 4, Part 9, Step 9	Discuss articles in *FOSS Science Resources* Inv 1, Part 4, Steps 7, 8, 12; Inv 1, Part 5, Steps 21, 22 Inv 2, Part 3, Steps 17, 18 Inv 3, Part 3, Steps 9, 10 Inv 4, Part 2, Steps 8, 9; Inv 4, Part 3, Steps 11, 12; Inv 4, Part 4, Steps 12, 14
Discuss and review articles in *FOSS Science Resources* Inv 1, Part 5, Steps 9, 10; Inv 1, Part 6, Step 2; Inv 1, Part 6, Step 11 Inv 2, Part 6, Step 7 Inv 3, Part 1, Steps 10, 11; Inv 3, Part 3, Step 10 Inv 4, Part 2, Step 8; Inv 4, Part 4, Step 9; Inv 4, Part 9, Step 9	Discuss and review articles in *FOSS Science Resources* Inv 1, Part 4, Step 11; Inv 1, Part 5, Step 22 Inv 2, Part 3, Steps 17, 18 Inv 3, Part 3, Steps 9, 10 Inv 4, Part 2, Steps 8, 9; Inv 4, Part 2, Steps 8, 9; Inv 4, Part 3, Steps 11, 12; Inv 4, Part 4, Steps 11, 12, 14
Discuss articles in *FOSS Science Resources* Inv 1, Part 5, Steps 9, 10; Inv 1, Part 6, Step 11 Inv 2, Part 6, Steps 6, 7 Inv 3, Part 1, Steps 10, 11; Inv 3, Part 3, Steps 9, 10, 12 Inv 4, Part 2, Step 8; Inv 4, Part 4, Step 9; Inv 4, Part 9, Step 9	Discuss articles in *FOSS Science Resources* Inv 1, Part 4, Steps 8, 11; Inv 1, Part 5, Steps 21, 22 Inv 2, Part 3, Steps 17, 18 Inv 3, Part 3, Steps 9, 10 Inv 4, Part 2, Steps 8-10; Inv 4, Part 3, Step 12; Inv 4, Part 4, Steps 12, 14
All investigations provide opportunities for students to ask and answer questions to determine the meaning of new words and phrases while reading articles in *FOSS Science Resources.* Selected examples Inv 2, Part 1, Step 1 Inv 3, Part 3, Step 12	All investigations provide opportunities for students to ask and answer questions to determine the meaning of new words and phrases while reading articles in *FOSS Science Resources.* Selected examples Inv 1, Part 4, Step 7 Inv 3, Part 3, Step 11
Discuss text features in *FOSS Science Resources* Inv 1, Part 5, Step 8 Inv 2, Part 1, Step 1	Discuss text features in *FOSS Science Resources* Inv 1, Part 4, Step 6

READING STANDARDS FOR INFORMATIONAL TEXT (CONT.)

	Grade K Standard	Materials and Motion Module
Integration of Knowledge and Ideas	7. With prompting and support, describe the relationship between illustrations and the text in which they appear (e.g., what person, place, thing, or idea in the text an illustration depicts).	All investigations provide opportunities for students to describe the relationship between the photographs and the text in the *FOSS Science Resources* articles. Selected examples Inv 1, Part 1, Steps 19, 20; Inv 1, Part 7, Step 9 Inv 2, Part 1, Steps 12, 13 Inv 3, Part 4, Step 8; Inv 3, Part 5, Step 1 Inv 4, Part 1, Steps 16, 17 Inv 4, Part 2, Steps 14, 15
	8. With prompting and support, identify the reasons an author gives to support points in a text.	Read and discuss articles in *FOSS Science Resources* Inv 1, Part 1, Steps 19, 20 Inv 3, Part 4, Step 8 Inv 4, Part 1, Step 17
	9. With prompting and support, identify basic similarities in and differences between two texts on the same topic (e.g., in illustrations, descriptions, or procedures).	Students can read their *FOSS Science Resources* as well as readings suggested on FOSSweb to compare and contrast how science ideas are communicated. Selected examples Inv 2, Part 1, Step 15; Inv 2, Art and Engineering Extensions. Examine paper-illustration techniques Inv 3, Science and Engineering Extensions, Show how knit fabric is made
Range of Reading and Level of Text Complexity	10. Actively engage in group reading activities with purpose and understanding.	All investigations involve interactive reading of science text. Selected examples Inv 1, Part 1, Steps 19, 20; Inv 1, Part 7, Steps 9, 10 Inv 2, Part 1, Steps 12-15 Inv 3, Part 2, Steps 10, 11; Inv 3, Part 4, Steps 8, 9; Inv 3, Part 5, Steps 1, 2 Inv 4, Part 1, Step 16 Inv 4, Part 2, Step 14

Trees and Weather Module	Animals Two by Two Module
All investigations provide opportunities for students to describe the relationship between the photographs and the text in the *FOSS Science Resources* articles. Selected examples Inv 1, Part 5, Steps 8, 9; Inv 1, Part 6, Step 11 Inv 2, Part 1, Step 1 Inv 3, Part 3, Step 12	All investigations provide opportunities for students to describe the relationship between the photographs and the text in the *FOSS Science Resources* articles. Selected examples Inv 1, Part 4, Steps 6, 9, 12; Inv 1, Part 5, Steps 21, 23 Inv 2, Part 3, Steps 17-20 Inv 3, Part 3, Step 12 Inv 4, Part 2, Step 11; Inv 4, Part 4, Step 14
Read and discuss articles in *FOSS Science Resources* Inv 4, Part 2, Step 8	Read and discuss articles in *FOSS Science Resources* Inv 1, Part 5, Step 21 Inv 2, Part 3, Step 20 Inv 3, Part 3, Step 12 Inv 4, Part 2, Step 11; Inv 4, Part 4, Step 14
Students can read their *FOSS Science Resources* as well as readings suggested on FOSSweb to compare and contrast how science ideas are communicated. Selected examples Inv 3, Part 3, Step 12; Inv 3, Language Extensions. Read weather related literature and poetry Inv 4, Part 2, Step 8	Students can read their *FOSS Science Resources* as well as readings suggested on FOSSweb to compare and contrast how science ideas are communicated. Selected examples Inv 2, Part 3, Step 20 Inv 3, Part 3, Step 12 Inv 4, Part 2, Step 11
All investigations involve interactive reading of science text. Selected examples Inv 1, Part 5, Steps 8-10; Inv 1, Part 6, Steps 1, 10, 11 Inv 2, Part 1, Step 1; Inv 2, Part 6, Steps 6, 7 Inv 3, Part 1, Steps 10, 11; Inv 3, Part 3, Steps 9-12 Inv 4, Part 2, Steps 7, 8; Inv 4, Part 4, Step 9; Inv 4, Part 9, Steps 8, 9	All investigations involve interactive reading of science text. Selected examples Inv 1, Part 4, Steps 6-12; Inv 1, Part 5, Steps 21-23 Inv 2, Part 3, Steps 17-20 Inv 3, Part 3, Steps 9-12 Inv 4, Part 2, Steps 8-11; Inv 4, Part 3, Steps 11-13; Inv 4, Part 4, Steps 11, 12, 14

READING STANDARDS: FOUNDATIONAL SKILLS

	Grade K Standard	Materials and Motion Module
Print Concepts	1. Demonstrate understanding of the organization and basic features of print. a. Follow words from left to right, top to bottom, and page by page. b. Recognize that spoken words are represented in written language by specific sequences of letters. c. Understand that words are separated by spaces in print. d. Recognize and name all upper- and lowercase letters of the alphabet.	All investigations provide opportunities for students to demonstrate understanding of the organization and basic features of print in *FOSS Science Resources*. Selected example Inv 3, Part 2, Step 12
Phonological Awareness	2. Demonstrate understanding of spoken words, syllables, and sounds (phonemes). a. Recognize and produce rhyming words. b. Count, pronounce, blend, and segment syllables in spoken words. c. Blend and segment onsets and rimes of single-syllable spoken words. d. Isolate and pronounce the initial, medial vowel, and final sounds (phonemes) in three-phoneme (consonant-vowel-consonant, or CVC) words.* (This does not include CVCs ending with /l/, /r/, or /x/.) e. Add or substitute individual sounds (phonemes) in simple, one-syllable words to make new words.	All investigations provide opportunities for students to demonstrate understanding of spoken words, syllables, and sounds while reading articles in *FOSS Science Resources*. Selected examples Inv 1, Part 1, Step 1; Inv 1, Part 2, Step 1; Inv 1, Part 3, Step 1; Inv 1, Part 5, Step 6

Trees and Weather Module	Animals Two by Two Module
All investigations provide opportunities for students to demonstrate understanding of the organization and basic features of print in *FOSS Science Resources*. Selected example Inv 4, Part 2, Step 8	All investigations provide opportunities for students to demonstrate understanding of the organization and basic features of print in *FOSS Science Resources*. Selected example Inv 4, Part 4, Step 14
All investigations provide opportunities for students to demonstrate understanding of spoken words, syllables, and sounds while reading articles in *FOSS Science Resources*. Selected examples Inv 1, Part 1, Step 1; Inv 1, Part 2, Step 5; Inv 1, Part 3, Step 1	All investigations provide opportunities for students to demonstrate understanding of spoken words, syllables, and sounds while reading articles in *FOSS Science Resources*.

READING STANDARDS: FOUNDATIONAL SKILLS (CON

	Grade K Standard	Materials and Motion Module
Phonics and Word Recognition	3. Know and apply grade-level phonics and word analysis skills in decoding words. a. Demonstrate basic knowledge of one-to-one letter-sound correspondences by producing the primary sound or many of the most frequent sounds for each consonant. b. Associate the long and short sounds with common spellings (graphemes) for the five major vowels. c. Read common high-frequency words by sight (e.g., *the, of, to, you, she, my, is, are, do, does*). d. Distinguish between similarly spelled words by identifying the sounds of the letters that differ.	All investigations provide opportunities for students to apply phonics and word analysis skills in decoding words while reading articles in *FOSS Science Resources*.
Fluency	4. Read emergent-reader texts with purpose and understanding.	All investigations provide opportunities for students to reread the articles in the *FOSS Science Resources* with a partner. Selected examples Inv 1, Part 1, Step 20 Inv 2, Part 1, Step 15 Inv 3, Part 2, Step 11; Inv 3, Part 4, Step 9; Inv 3, Part 5, Step 2

Trees and Weather Module	Animals Two by Two Module
All investigations provide opportunities for students to apply phonics and word analysis skills in decoding words while reading articles in *FOSS Science Resources*.	All investigations provide opportunities for students to apply phonics and word analysis skills in decoding words while reading articles in *FOSS Science Resources*.
All investigations provide opportunities for students to reread the articles in the *FOSS Science Resources* with a partner. Selected examples Inv 1, Part 5, Step 10; Inv 1, Part 6, Step 11 Inv 3, Part 1, Steps 11; Inv 3, Part 3, Step 10 Inv 4, Part 2, Step 8; Inv 4, Part 4, Step 9; Inv 4, Part 9, Steps 8, 9	All investigations provide opportunities for students to reread the articles in the *FOSS Science Resources* with a partner. Selected examples Inv 4, Part 3, Step 13; Inv 4, Part 4, Step 14

WRITING STANDARDS

	Grade K Standard	Materials and Motion Module
Text Types and Purposes	1. Use a combination of drawing, dictating, and writing to compose opinion pieces in which they tell a reader the topic or the name of the book they are writing about and state an opinion or preference about the topic or book (e.g., *My favorite book is . . .*).	All investigations provide opportunities for students to draw, dictate, or write about a science topic, stating their opinion or claim. Selected examples Inv 1, Part 1, Step 20; Inv 1, Part 5, Step 16 Inv 2, Part 2, Step 18 Inv 3, Part 5, Step 9
Text Types and Purposes	2. Use a combination of drawing, dictating, and writing to compose informative/explanatory texts in which they name what they are writing about and supply some information about the topic.	All investigations provide opportunities for students to draw, dictate, or write informative/explanatory texts about the science topic they are learning. Selected examples Inv 1, Part 1, Steps 17, 21; Inv 1, Part 2, Step 17; Inv 1, Part 4, Step 11; Inv 1, Language Extensions, Create a class book of objects make of wood. Inv 2, Part 1, Step 10 Inv 3, Part 1, Step 17; Inv 3, Part 2, Step 9; Inv 3, Part 3, Step 10; Inv 3, Part 4, Step 10 Inv 4, Part 1, Step 15; Inv 4, Part 2, Step 13
Text Types and Purposes	3. Use a combination of drawing, dictating, and writing to narrate a single event or several loosely linked events, tell about the events in the order in which they occurred, and provide a reaction to what happened.	All investigations provide opportunities for students to draw, dictate, and write narratives. Students describe their observations and experiences with the science ideas they are exploring. Selected examples Inv 1, Part 1, Step 21; Inv 1, Part 3, Step 21; Inv 1, Part 6, Step 13; Inv 1, Part 7, Step 8 Inv 2, Part 3, Step 7; Inv 2, Part 4, Step 10; Inv 2, Part 5, Step 14 Inv 4, Part 3, Step 10; Inv 4, Language exension
Production and Distribution of Writing	5. With guidance and support from adults, respond to questions and suggestions from peers and add details to strengthen writing as needed.	The Wrap-up/Warm-up section of each investigation part, provides the opportunity for students to strengthen their notebook entries by revising and adding in new information. Selected examples Inv 1, Part 3, Step 22; Inv 1, Part 5, Step 17 Inv 2, Part 2, Step 19 Inv 4, Part 2, Step 16; Inv 4, Part 4, Step 11

Trees and Weather Module	Animals Two by Two Module
All investigations provide opportunities for students to draw, dictate, and write about a science topic, stating their opinion or claim. Selected examples Inv 3, Part 1, Step 8 Inv 4, Part 9, Step 6	All investigations provide opportunities for students to draw, dictate, and write about a science topic, stating their opinion or claim. Selected example Inv 2, Part 2, Step 9
All investigations provide opportunities for students to draw, dictate, or write informative/explanatory texts about the science topic they are learning. Selected examples Inv 1, Part 1, Step 13; Inv 1, Part 3, Step 6; Inv 1, Part 4, Step 4; Inv 1, Part 6, Step 9 Inv 2, Part 1, Step 13; Inv 2, Part 2, Step 9; Inv 2, Part 3, Step 9 Inv 3, Part 3, Step 6	All investigations provide opportunities for students to draw, dictate, or write informative/explanatory texts about the science topic they are learning. Selected examples Inv 1, Part 1, Step 4; Inv 1, Part 2, Step 8; Inv 1, Part 3, Step 6; Inv 1, Part 4, Step 4; Inv 1, Language Extensions, Write a "facts about fish" book Inv 2, Part 1, Step 6; Inv 2, Part 2, Step 9 Inv 3, Part 1, Step 8; Inv 3, Part 2, Step 16 Inv 4, Part 1, Step 9; Inv 4, Part 2, Step 6; Inv 4, Part 3, Step 8
All investigations provide opportunities for students to draw, dictate, and write narratives. Students describe their observations and experiences with the science ideas they are exploring. Selected examples Inv 1, Part 5, Step 6 Inv 2, Part 5, Step 4 Inv 3, Part 2, Step 13 Inv 4, Part 3, Step 5; Inv 4, Part 6, Step 6	All investigations provide opportunities for students to draw, dictate, and write narratives. Students describe their observations and experiences with the science ideas they are exploring. Selected examples Inv 1, Part 5, Steps 13, 20; Inv 1, Language Extensions, Write a story Inv 2, Part 3, Steps 12, 13; Inv 2, Language Extensions, Keep a classroom snail journal
The Wrap-up/Warm-up section of each investigation part, provides the opportunity for students to strengthen their notebook entries by revising and adding in new information. Selected examples Inv 1, Part 5, Step 7 Inv 2, Part 1, Step 14; Inv 2, Part 3, Step 11 Inv 3, Part 3, Step 7 Inv 4, Part 3, Step 7; Inv 4, Part 6, Step 8; Inv 4, Part 9, Step 7	The Wrap-up/Warm-up section of each investigation part, provides the opportunity for students to strengthen their notebook entries by revising and adding in new information. Selected examples Inv 1, Part 4, Step 13 Inv 2, Part 1, Step 11 Inv 3, Part 1, Step 11 Inv 4, Part 2, Step 12

WRITING STANDARDS (CONT.)

	Grade K Standard	Materials and Motion Module
Research to Build and Present Knowledge	7. Participate in shared research and writing projects (e.g., explore a number of books by a favorite author and express opinions about them).	In every investigation students record their observations in their notebooks. Selected examples Inv 1 Part 1 Steps 18, 21; Inv 1, Part 2, Step 17; Inv 1, Part 3, Step 22; Inv 1, Part 4, Step 11; Inv 1, Part 5, Step 16; Inv 1, Part 6, Step 13; Inv 1, Part 7, Step 8; Inv 1, Language Extensions, Create a wood chart, Create a class book of objects make of wood. Inv 2, Part 1, Step 10; Inv 2, Part 2, Step 18; Inv 2, Part 3, Step 7; Inv 2, Part 5, Step 14; Inv 2, Language Extension, Make a paper chart Inv 3, Part 1, Step 17; Inv 3, Part 2, Steps 9, 12; Inv 3, Part 3, Step 10; Inv 3, Part 4, Step 10; Inv 3, Part 5, Step 9
	8. With guidance and support from adults, recall information from experiences or gather information from provided sources to answer a question.	All investigations provide students with the opportunity to write about their science experiences and record their observations in their science notebooks. Selected examples Inv 1, Part 1, Step 17, Inv 1, Part 2, Step 17; Inv 1, Part 3, Steps 12, 21; Inv 1, Part 4, Step 11; Inv 1, Part 5, Step 16; Inv 1, Part 6, Step 13; Inv 1, Part 7, Step 8 Inv 2, Part 1, Step 10; Inv 2, Part 2, Step 18; Inv 2, Part 3, Step 7; Inv 2, Part 5, Step 14 Inv 3, Part 1, Step 17; Inv 3, Part 2, Steps 9, 12; Inv 3, Part 3, Step 10; Inv 3, Part 4, Step 10; Inv 3, Part 5, Step 9 Inv 4, Part 1, Step 15; Inv 4, Part 2, Step 13; Inv 4, Part 3, Step 10

Trees and Weather Module	Animals Two by Two Module
In every investigation students record their observations in their notebooks. Selected examples Inv 1, Part 1, Step 13; Inv 1, Part 3, Step 6; Inv 1, Part 6, Steps 9, 10; Language Extension. Make a tree-observation class book Inv 2, Part 1, Step 13; Inv 2, Part 2, Step 9; Inv 2, Part 3, Step 9 Inv 3, Part 1, Step 8; Inv 3, Part 2, Step 13; Inv 3, Part 3, Step 6 Inv 4, Part 3, Step 5; Inv 4, Part 3, Step 7; Inv 4, Part 5, Step 7; Inv 4, Part 9, Step 4	In every investigation students record their observations in their notebooks. Selected examples Inv 1, Part 1, Steps 4, 5; Inv 1, Part 2, Step 8; Inv 1, Part 3, Step 6; Inv 1, Part 4, Steps 4, 12; Inv 1, Part 5, Steps 13, 20; Inv 1, Science Extensions, Identify animals that eat fish Inv 2, Part 1, Step 6; Inv 2, Part 3, Steps 12, 13, 16 Inv 3, Part 1, Step 8; Inv 3, Part 2, Step 16; Language Extension, Keep a classroom worm book Inv 4, Part 1, Step 9; Inv 4, Language Extension, Make a classroom isopod journal; Inv 4, Science Extension, Find out about other crustaceans
All investigations provide students with the opportunity to write about their science experiences and record their observations in their science notebooks. Selected examples Inv 1, Part 1, Step 13; Inv 1, Part 3, Step 6; Inv 1, Part 5, Step 6; Inv 1, Part 6, Step 9 Inv 2, Part 1, Step 13; Inv 2, Part 2, Step 9; Inv 2, Part 3, Step 9 Inv 3, Part 1, Step 8; Inv 3, Part 2, Step 13; Inv 3, Part 3, Step 6 Inv 4, Part 2, Step 8; Inv 4, Part 3, Step 5; Inv 4, Part 6, Step 6	All investigations provide students with the opportunity to write about their science experiences and record their observations in their science notebooks. Selected examples Inv 1, Part 1, Step 4; Inv 1, Part 2, Step 8; Inv 1, Part 3, Step 6; Inv 1, Part 4, Step 4; Inv 1, Part 5, Steps 13, 20, 22 Inv 2, Part 1, Step 6; Inv 2, Part 3, Steps 12, 13 Inv 3, Part 1, Step 8; Inv 3, Part 2, Step 16 Inv 4, Part 1, Step 9

SPEAKING AND LISTENING STANDARDS

Comprehension and Collaboration

Grade K Standard	Materials and Motion Module
1. Participate in collaborative conversations with diverse partners about *kindergarten topics and texts* with peers and adults in small and larger groups. a. Follow agreed-upon rules for discussions (e.g., listening to others and taking turns speaking about the topics and texts under discussion). b. Continue a conversation through multiple exchanges.	All investigations provide students ample opportunities to engage in collaborative discussions. Students discuss before, during, and after the active investigation, when reading articles in the *FOSS Science Resources*, and during the Wrap-up/Warm-up section. Selected examples Inv 1, Part 1, Steps 11, 19, 20, 22; Inv 1, Part 2, Steps 13, 18; Inv 1, Part 3, Step 22; Inv 1, Part 4, Step 12; Inv 1, Part 5, Steps 1, 17; Inv 1, Part 6, Step 15 Inv 2, Part 1, Step 11; Inv 2, Part 2, Steps 2, 19; Inv 2, Part 3, Step 15; Inv 2, Part 4, Step 11 Inv 3, Part 1, Step 18; Inv 3, Part 2, Step 14; Inv 3, Part 3, Step 10; Inv 3, Part 4, Steps 8, 9, 11; Inv 3, Part 5, Step 14 Inv 4, Part 1, Step 18
2. Confirm understanding of a text read aloud or information presented orally or through other media by asking and answering questions about key details and requesting clarification if something is not understood.	All investigations provide opportunities for students to ask and answer questions about key details in *FOSS Science Resources* articles and information presented orally. Selected examples Inv 1, Part 1, Steps 19, 20, 22; Inv 1, Part 2, Steps 6, 10, 15; Inv 1, Part 3, Steps 8, 22; Inv 1, Part 4, Steps 3, 5; Inv 1, Part 6, Steps 1, 4, 13; Inv 1, Part 7, Step 10 Inv 2, Part 2, Step 7 Inv 3, Part 2, Step 5; Inv 3, Part 3, Steps 1,2; Inv 3, Part 4, Steps 8, 9 Inv 4, Part 1, Step 17
3. Ask and answer questions in order to seek help, get information, or clarify something that is not understood.	All investigations provide students with the opportunity to ask and answer questions about how they answered the focus question during the Wrap-up/Warm-up section. Other opportunities arise when students present information to their group or the whole class. Selected examples Inv 1, Part 1, Steps 4, 6; Inv 1, Part 2, Step 2; Inv 1, Part 4, Step 8; Inv 1, Part 5, Steps 8, 10, 12, 13 Inv 2, Part 2, Steps 5, 14; Inv 2, Part 3, Steps 1, 3, 4; Inv 2, Part 4, Step 6; Inv 2, Part 5, Steps 6, 11 Inv 3, Part 1, Steps 4, 8; Inv 3, Part 2, Step 1; Inv 3, Part 3, Steps 5-7; Inv 3, Part 4, Steps 5, 7; Inv 3, Part 5, Step 1

Trees and Weather Module	Animals Two by Two Module
All investigations provide students ample opportunities to engage in collaborative discussions. Students discuss before, during, and after the active investigation, when reading articles in the *FOSS Science Resources*, and during the Wrap-up/Warm-up section. Selected examples Inv 1, Part 1, Steps 15, 16; Inv 1, Part 3, Step 7; Inv 1, Part 4, Step 6; Inv 1, Part 6, Step 12 Inv 2, Part 1, Step 14; Inv 2, Part 2, Step 10; Inv 2, Part 3, Step 11; Inv 2, Part 5, Step 1 Inv 3, Part 1, Step 9; Inv 3, Part 2, Step 14; Inv 3, Part 3, Step 7 Inv 4, Part 6, Step 8; Inv 4, Part 9, Step 7	All investigations provide students ample opportunities to engage in collaborative discussions. Students discuss before, during, and after the active investigation, when reading articles in the *FOSS Science Resources*, and during the Wrap-up/Warm-up section. Selected examples Inv 1, Part 1, Step 6; Inv 1, Part 2, Step 10; Inv 1, Part 3, Step 8; Inv 1, Part 5, Step 24 Inv 2, Part 2, Step 11; Inv 2, Part 3, Steps 21, 22 Inv 3, Part 1, Step 11; Inv 3, Part 2, Step 19; Inv 3, Part 3, Step 13 Inv 4, Part 1, Step 12; Inv 4, Part 2, Step 12; Inv 4, Part 3, Step 14
All investigations provide opportunities for students to ask and answer questions about key details in *FOSS Science Resources* articles and information presented orally. Selected examples Inv 1, Part 1, Step 10; Inv 1, Part 4, Step 6; Inv 1, Part 5, Steps 1, 3, 6, 10; Inv 1, Part 6, Step 12 Inv 2, Part 2, Step 1; Inv 2, Part 6, Steps 6, 7 Inv 3, Part 1, Steps 10, 11; Inv 3, Part 2, Step 7; Inv 3, Part 3, Steps 8, 10, 12 Inv 4, Part 5, Step 1; Inv 4, Part 9, Step 9	All investigations provide opportunities for students to ask and answer questions about key details in *FOSS Science Resources* articles and information presented orally. Selected examples Inv 1, Part 3, Steps 1, 2, 4; Inv 1, Part 4, Steps 8, 11; Inv 1, Part 5, Steps 7, 14 Inv 2, Part 1, Step 10; Inv 2, Part 2, Step 5; Inv 2, Part 3, Step 15 Inv 3, Part 2, Steps 8, 14; Inv 3, Part 3, Steps 5, 13 Inv 4, Part 3, Steps 6, 14; Inv 4, Part 4, Step 2
All investigations provide students with the opportunity to ask and answer questions about how they answered the focus question during the Wrap-up/Warm-up section. Other opportunities arise when students present information to their group or the whole class. Selected examples Inv 1, Part 1, Steps 4-6; Inv 1, Part 5, Step 10; Inv 1, Part 6, Step 2 Inv 2, Part 1, Step 3; Inv 2, Part 4, Step 6 Inv 3, Part 2, Steps 9-11; Inv 3, Part 3, Steps 1, 3 Inv 4, Part 1, Steps 1, 2; Inv 4, Part 2, Step 1; Inv 4, Part 5, Steps 3, 6; Inv 4, Part 7, Step 1; Inv 4, Part 8, Steps 1, 2	All investigations provide students with the opportunity to ask and answer questions about how they answered the focus question during the Wrap-up/Warm-up section. Other opportunities arise when students present information to their group or the whole class. Selected examples Inv 1, Part 1, Step 2; Inv 1, Part 2, Steps 3-6; Inv 1, Part 4, Steps 2, 8, 11; Inv 1, Part 5, Steps 3, 5, 6, 10, 11, 16, 20 Inv 2, Part 1, Steps 3-5; Inv 2, Part 2, Step 6; Inv 2, Part 3, Steps 5, 7, 11 Inv 3, Part 1, Step 6; Inv 3, Part 2, Steps 4-6, 18; Inv 3, Part 3, Steps 3, 4 Inv 4, Part 1, Step 4; Inv 4, Part 3, Steps 3-5; Inv 4, Part 4, Step 6

SPEAKING AND LISTENING STANDARDS (CONT.)

	Grade K Standard	Materials and Motion Module
Presentation of Knowledge and Ideas	4. Describe familiar people, places, things, and events and, with prompting and support, provide additional detail.	All investigations provide students with the opportunity to describe what they observe and learn about science topics. In the Wrap-up/Warm-up section students describe what they did in the investigation and share their answers to the focus question. Selected examples Inv 1, Part 1, Steps 6, 14, 19; Inv 1, Part 3, Step 13; Inv 1, Part 4, Step 1; Inv 1, Part 5, Step 1; Inv 1, Part 6, Step 1 Inv 3, Part 1, Step 1 Inv 4, Part 1, Steps 16, 17
	5. Add drawings or other visual displays to descriptions as desired to provide additional detail.	Students are encouraged to use their notebook drawings to explain their answers to the focus question in the Wrap-up/Warm-up sections. Selected examples Inv 1, Part 1, Step 22; Inv 1, Part 3, Steps 18, 22; Inv 1, Part 4, Step 12; Inv 1, Part 5, Step 17; Inv 1, Part 6, Step 15; Inv 1, Part 7, Step 11 Inv 3, Part 1, Step 18; Inv 3, Part 5, Step 14; Inv 3, Part 6, Step 1 Inv 4, Part 1, Step 18 Inv 4, Part 2, Step 16
	6. Speak audibly and express thoughts, feelings, and ideas clearly.	All investigations provide students with the opportunity to speak audibly and express their thoughts, feelings, and ideas about their science learning. Selected examples Inv 1, Part 1, Steps 14, 19, 20; Inv 1, Part 3, Steps 8, 22 Inv 2, Part 4, Step 2 Inv 3, Part 1, Step 18; Inv 3, Part 2, Step 12; Inv 3, Part 3, Step 3; Inv 3, Part 4, Steps 9, 11; Inv 3, Part 5, Step 7 Inv 4, Part 2, Step 16

Trees and Weather Module	Animals Two by Two Module
All investigations provide students with the opportunity to describe what they observe and learn about science topics. In the Wrap-up/Warm-up section students describe what they did in the investigation and share their answers to the focus question. Selected examples Inv 1, Part 1, Step 2; Inv 1, Part 5, Steps 7, 9, 10, 11; Inv 1, Part 6, Steps 10, 12 Inv 2, Part 6, Steps 6, 7 Inv 3, Part 1, Steps 2, 3, 10; Inv 3, Part 2, Steps 4, 5 Inv 4, Part 2, Step 8; Inv 4, Part 4, Steps 8, 9; Inv 4, Part 6, Step 1; Inv 4, Part 7, Step 1; Inv 4, Part 9, Step 1	All investigations provide students with the opportunity to describe what they observe and learn about science topics. In the Wrap-up/Warm-up section students describe what they did in the investigation and share their answers to the focus question. Selected examples Inv 1, Part 4, Step 6; Inv 1, Part 5, Step 1 Inv 2, Part 1, Step 2; Inv 2, Part 2, Step 2; Inv 2, Part 3, Step 1 Inv 4, Part 1, Step 1; Inv 4, Part 3, Step 11
Students are encouraged to use their notebook drawings to explain their answers to the focus question in the Wrap-up/Warm-up sections. Selected examples Inv 1, Part 2, Step 5; Inv 1, Part 4, Step 6; Inv 1, Part 5, Step 11 Inv 3, Part 1, Step 11; Inv 3, Part 2, Step 14; Inv 3, Part 3, Step 7 Inv 4, Part 9, Step 7	Students are encouraged to use their notebook drawings to explain their answers to the focus question in the Wrap-up/Warm-up sections. Selected examples Inv 1, Part 3, Step 4 Inv 3, Part 1, Step 11; Inv 3, Part 2, Step 19 Inv 4, Part 2, Step 12; Inv 4, Part 4, Step 8
All investigations provide students with the opportunity to speak audibly and express their thoughts, feelings, and ideas about their science learning. Selected examples Inv 1, Part 1, Step; Inv 1, Part 3, Step 7; Inv 1, Part 4, Step 6; Inv 1, Part 6, Step 12 Inv 2, Part 6, Steps 6, 7 Inv 3, Part 1, Step 9; Inv 3, Part 2, Step 1; Inv 3, Part 3, Steps 7, 12 Inv 4, Part 2, Step 7; Inv 4, Part 7, Step 2; Inv 4, Part 9, Step 9	All investigations provide students with the opportunity to speak audibly and express their thoughts, feelings, and ideas about their science learning. Selected examples Inv 1, Part 3, Step 1 Inv 2, Part 3, Step 22 Inv 3, Part 3, Step 13 Inv 4, Part 3, Step 1; Inv 4, Part 3, Step 12

LANGUAGE STANDARDS

	Grade K Standard	Materials and Motion Module
Conventions of Standard English	1. Demonstrate command of the conventions of standard English grammar and usage when writing or speaking. a. Print many upper- and lowercase letters. b. Use frequently occurring nouns and verbs. c. Form regular plural nouns orally by adding /s/ or /es/ (e.g., *dog, dogs; wish, wishes*). d. Understand and use question words (interrogatives) (e.g., *who, what, where, when, why, how*). e. Use the most frequently occurring prepositions (e.g., *to, from, in, out, on, off, for, of, by, with*). f. Produce and expand complete sentences in shared language activities.	All investigations provide opportunities for students to demonstrate the conventions of standard English grammar when writing and speaking. Selected examples Inv 1, Part 1, Steps 1, 5, 14, 15, 17, 18, 21; Inv 1, Part 2, Step 17; Inv 1, Part 3, Steps 11, 12; Inv 1, Part 4, Steps 10-12; Inv 1, Part 5, Step 16; Inv 1, Part 7, Step 8 Inv 2, Part 1, Step 10 Inv 3, Part 2, Step 12
	2. Demonstrate command of the conventions of standard English capitalization, punctuation, and spelling when writing. a. Capitalize the first word in a sentence and the pronoun *I*. b. Recognize and name end punctuation. c. Write a letter or letters for most consonant and short-vowel sounds (phonemes). d. Spell simple words phonetically, drawing on knowledge of sound-letter relationships.	All investigations provide opportunities for students to demonstrate command of the conventions of standard English capitalization, punctuation, and spelling when writing in their science notebooks.
Vocabulary Acquisition and Use	4. Determine or clarify the meaning of unknown and multiple-meaning words and phrases based on *kindergarten reading and content*. a. Identify new meanings for familiar words and apply them accurately (e.g., knowing *duck* is a bird and learning the verb to *duck*). b. Use the most frequently occurring inflections and affixes (e.g., *-ed, -s, re-, un-, pre-, -ful, -less*) as a clue to the meaning of an unknown word.	All investigations provide opportunities for students to practice strategies for determining or clarifying the meaning of unknown and multiple-meaning words and phrases while discussing the investigations and articles in *FOSS Science Resources*. Selected examples Inv 1, Part 1, Step 15; Inv 1, Part 2, Step 12; Inv 1, Part 3, Step 19; Inv 1, Part 7, Step 7 Inv 2, Part 1, Step 9; Inv 2, Part 2, Step 17; Inv 2, Part 3, Step 6; Inv 2, Part 4, Step 9; Inv 2, Part 5, Step 12 Inv 3, Part 1, Step 14; Inv 3, Part 2, Step 12; Inv 3, Part 3, Step 9; Inv 3, Part 5, Steps 2, 3, 5

Trees and Weather Module	Animals Two by Two Module
All investigations provide opportunities for students to apply the conventions of English grammar when writing and speaking. Selected examples Inv 1, Part 6, Step 9 Inv 2, Part 1, Step 13; Inv 2, Part 2, Step 9; Inv 2, Part 3, Step 9 Inv 3, Part 1, Steps 9, 10; Inv 3, Part 2, Step 13; Inv 3, Part 3, Steps 6, 7	All investigations provide opportunities for students to apply the conventions of English grammar when writing and speaking. Selected examples Inv 1, Part 2, Step 8; Inv 1, Part 3, Step 6; Inv 1, Part 5, Step 20 Inv 2, Part 2, Step 9; Inv 2, Part 3, Step 22 Inv 3, Part 2, Step 16
All investigations provide opportunities for students to demonstrate command of the conventions of standard English capitalization, punctuation, and spelling when writing in their science notebooks.	All investigations provide opportunities for students to demonstrate command of the conventions of standard English capitalization, punctuation, and spelling when writing in their science notebooks.
All investigations provide opportunities for students to practice strategies for determining or clarifying the meaning of unknown and multiple-meaning words and phrases while discussing the investigations and articles in *FOSS Science Resources.* Selected examples Inv 1, Part 1, Step 12; Inv 1; Inv 1, Part 5, Step 5 Inv 2, Part 3, Step 8; Inv 2, Part 4, Step 7 Inv 3, Part 1, Step 4; Inv 3, Part 2, Step 12; Inv 3, Part 3, Step 5 Inv 4, Part 3, Step 4; Inv 4, Part 6, Step 5	All investigations provide opportunities for students to practice strategies for determining or clarifying the meaning of unknown and multiple-meaning words and phrases while discussing the investigations and articles in *FOSS Science Resources.* Selected examples Inv 1, Part 1, Step 3; Inv 1, Part 2, Step 7; Inv 1, Part 3, Step 5; Inv 1, Part 4, Step 3; Inv 1, Part 5, Steps 12, 19 Inv 2, Part 1, Step 9; Inv 2, Part 2, Step 8; Inv 2, Part 3, Step 9 Inv 3, Part 1, Step 7; Inv 3, Part 2, Step 15; Inv 3, Part 3, Step 6 Inv 4, Part 1, Step 8

LANGUAGE STANDARDS (CONT.)

Vocabulary Acquisition and Use

Grade K Standard	Materials and Motion Module
5. With guidance and support from adults, explore word relationships and nuances in word meanings. a. Sort common objects into categories (e.g., shapes, foods) to gain a sense of the concepts the categories represent. b. Demonstrate understanding of frequently occurring verbs and adjectives by relating them to their opposites (antonyms). c. Identify real-life connections between words and their use (e.g., note places at school that are *colorful*). d. Distinguish shades of meaning among verbs describing the same general action (e.g., *walk, march, strut, prance*) by acting out the meanings.	All investigations provide students with opportunities to explore word relationships and nuances of certain words that have a specific meaning in science, such as **observe, senses, material, rough, smooth, grain, properties, absorb, bead, float, sink, soak, spread, graph, change, mixture, shavings, waterlogged, separate, screen, evaporate, mix, layers, paper, fabric, cloth, waterproof,** and **structures.** Selected examples Inv 1, Part 1, Step 4; Inv 1, Part 2, Steps 6, 10; Inv 1, Part 3, Steps 2, 7; Inv 1, Part 4, Steps 1, 5; Inv 1, Part 5, Steps 6, 16; Inv 1, Part 6, Steps 1, 2; Inv 1, Part 7, Steps 1, 7; Inv 1, Language Extension, Create a sorting challenge Inv 2, Part 1, Steps 3, 9; Inv 2, Part 2, Step 8 Inv 3, Part 1, Step 9; Inv 3, Part 2, Steps 3, 5; Inv 3, Part 3, Step 4; Inv 3, Language Extension, Make word and fabric cards
6. Use words and phrases acquired through conversations, reading and being read to, and responding to texts.	All investigations provide opportunities for students to use new science words and phrases acquired through science discussions and readings. Science vocabulary words are in bold when they are first introduced to students in *FOSS Science Resources*. Students also review the vocabulary in the Review vocabulary section for each part of each investigation. Selected examples Inv 1, Part 1, Steps 1, 2, 4-9, 15, 17-18; Inv 1, Part 2, Steps 1, 6, 7, 10, 12, 17, 18; Inv 1, Part 3, Steps 1, 4, 7, 9, 11, 12, 17, 18, 19, 21; Inv 1, Part 4, Steps 1, 2, 5, 10, 11; Inv 1, Part 5, Steps 6, 8-10, 15-18; Inv 1, Part 6, Steps 2, 5, 6, 9, 12, 13; Inv 1, Part 7, Steps 1, 5, 7, 8 Inv 2, Part 1, Steps 9, 10; Inv 2, Part 2, Steps 8, 14, 17, 18; Inv 2, Part 3, Step 6; Inv 2, Part 4, Steps 1, 3, 7, 9, 10; Inv 2, Part 5, Steps 1, 4, 7, 10, 12 Inv 3, Part 1, Steps 1, 5, 9, 12, 14, 17, 18; Inv 3, Part 3, Steps 3, 5, 8, 9, 12, 14; Inv 3, Part 5, Steps 2, 3, 5, 8, 9; Inv 3, Part 6, Step 1

Trees and Weather Module	Animals Two by Two Module
All investigations provide students with opportunities to explore word relationships and nuances of certain words that have a specific meaning in science, such as **observe, plant, tree, trunk, stem, roots, compare, shape, patterns, similar, adopt, living, nonliving, bark, circumference, textures, flowers, seeds, cones, need, space, toothed, lobed, rough, rounded, property, paddle. line, spear, oval, heart, silhouette, outline, weather, air, clouds, monitor, hot, cold, temperature, thermometer, freezing, cold, warm, wind, direction, blossom, bud, evergreen, fall, flower, food, forcing, needle scale, season, spring, summer, winter,** and **swollen.** Selected examples Inv 1, Part 3, Steps 2, 3, 6; Inv 1, Part 4, Steps 2, 5, 6; Inv 1, Part 5, Step 13; Inv 1, Part 6, Step 10 Inv 2, Part 1, Step 9; Inv 2, Part 2, Step 2; Inv 2, Part 3, Steps 4-6; Inv 2, Part 4, Step 6 Inv 3, Part 2, Step 1; Inv 3, Part 3, Step 10 Inv 4, Part 3, Step 2; Inv 4, Parts 1, 6; Inv 4, Part 7, Steps 1, 2	All investigations provide students with opportunities to explore word relationships and nuances of certain words that have a specific meaning in science, such as **animal, parts, surface, male, female, bill, foot, tentacle, shell, rough, smooth, soil, segment, shelter, bristle, isopods, section, antennae, protect, living,** and **nonliving.** Selected examples Inv 1, Part 2, Steps 3, 5, 6; Inv 1, Part 3, Step 4; Inv 1, Part 4, Steps 2, 12; Inv 1, Part 5, Step 14 Inv 2, Part 1, Step 10; Inv 2, Part 2, Steps 6, 7; Inv 2, Part 3, Steps 8, 16 Inv 3, Part 2, Step 4; Inv 3, Part 3, Steps 3, 13 Inv 4, Part 4, Step 12
All investigations provide opportunities for students to use new science words and phrases acquired through science discussions and readings. Science vocabulary words are in bold when they are first introduced to students in *FOSS Science Resources.* Students also review the vocabulary in the Review vocabulary section for each part of each investigation. Selected examples Inv 1, Part 1, Steps 1, 2, 4, 5, 6, 9; Part 2, Steps 3-5; Inv 1, Part 3, Steps 1, 2, 7; Inv 1, Part 4, Steps 5, 6; Inv 1, Part 5, Steps 2-5; Inv 1, Part 6, Step 11 Inv 2, Part 1, Steps 9, 10, 13, 14; Inv 2, Part 2, Steps 3-5, 9, 10; Inv 2, Part 3, Steps 4-6, 8, 9, 11; Inv 2, Part 4, Steps 2, 3, 7 Inv 3, Part 1, Steps 1, 2, 4-6, 8, 9; Inv 3, Part 2, Steps 1, 2, 12; Inv 3, Part 3, Steps 1, 2, 5, 6, 10 Inv 4, Part 2, Steps 3, 8; Inv 4, Part 3, Steps 1, 4; Inv 4, Parts 1, 4; Inv 4, Part 5, Steps 3, 4; Inv 4, Part 6, Step 5; Inv 4, Part 7, Steps 1, 2; Inv 4, Part 8, Step 5; Inv 4, Part 9, Steps 4, 5	All investigations provide opportunities for students to use new science words and phrases acquired through science discussions and readings. Science vocabulary words are in bold when they are first introduced to students in *FOSS Science Resources.* Students also review the vocabulary in the Review vocabulary section for each part of each investigation. Selected examples Inv 1, Part 1, Steps 1-4; Inv 1, Part 2, Steps 1, 3-8; Inv 1, Part 3, Steps 1, 2, 5; Inv 1, Part 4, Steps 2-4; Inv 1, Part 5, Steps 1, 7, 10, 12, 13, 19 Inv 2, Part 1, Steps 3, 9, 10; Inv 2, Part 2, Steps 2, 7-9; Inv 2, Part 3, Steps 6, 8, 9,11,12 Inv 3, Part 1, Steps 2, 3, 6, 7; Inv 3, Part 2, Steps 4, 14-16; Inv 3, Part 3, Steps 3, 4 Inv 4, Part 1, Steps 2, 4, 8; Inv 4, Part 2, Steps 1, 3, 5; Inv 4, Part 3, Step 8; Inv 4, Part 4, Step 7

FOSS and Common Core Math – Grade K

FOSS and Common Core Math — Grade K

Contents

INTRODUCTION

The adoption of the Common Core State Standards for Mathematics calls for shifts in focus, coherence, and rigor. The teaching of the standards should be focused on the important content, coherent from one grade level to the next, and rigorous in requiring conceptual understanding, fluency, and application. Within this area of application, FOSS provides fertile ground for the use of mathematics.

The FOSS Program integrates mathematics with science in two ways throughout the kindergarten modules. In active investigations, students apply mathematics during data gathering and analysis. In addition, the Interdisciplinary Extensions at the end of each investigation usually include a math problem of the week. These problems enhance the science learning by providing hypothetical data for students to analyze or in some way relate to the context of the investigation. The notes explain for the teacher the problem and describe how students might approach its solution. The problems are prepared for distribution to students on duplication masters in the Teacher Masters chapter of *Teacher Resources*.

This chapter gives an overview of how FOSS addresses the Common Core State Standards for Mathematics through science. It also points out specific instances in which students exercise those skills during science instruction.

Mathematical Practices

Mathematical practices consist of eight processes and proficiencies that are important for all students.

1. Make sense of problems and persevere in solving them.
2. Reason abstractly and quantitatively.
3. Construct viable arguments and critique the reasoning of others.
4. Model with mathematics.
5. Use appropriate tools strategically.
6. Attend to precision.
7. Look for and make use of structure.
8. Look for and express regularity in repeated reasoning.

Within the context of science, students use some of these mathematical practices on a regular basis. According to *Next Generation Science Standards* (volume 2, appendix L, p. 138),

The three CCSSM practice standards most directly relevant to science are:

- MP.2. Reason abstractly and quantitatively.
- MP.4. Model with mathematics.
- MP.5. Use appropriate tools strategically.

When students reason abstractly and quantitatively and model with mathematics, they are using math in context. They work with symbols and their meanings and represent and solve word problems. Students choose and correctly use the available tools to collect data and solve problems. In the kindergarten modules, students engage with these three practices during the active investigation and by completing the math extensions at the end of each investigation. Here are some examples.

In the **Animals Two by Two Module**, students count and answer questions about the number of different fish in a tank. In order to solve this problem, students reason quantitatively and use mathematics in the context of science. They count the fish and model using counters to compare the quanitities of the fish.

In the **Trees and Weather Module**, students make a list of items made of wood found in their home. The teacher guides students to sort these items into categories such as furniture and count them. The story problems provide opportunities for students to ulitilze tools to answer questions about the quanity of items.

In the **Materials and Motion Module**, students are asked to reason abstractly to determine the cost of caps. This requires students to count the number of caps and then add the cost of each cap. The teacher can modify the cost of the caps to challenge students.

Mathematical Content

The mathematical content in kindergarten is organized around five concepts.

- Counting and Cardinality
- Operations and algebraic thinking
- Measurement and data
- Geometry

The following pages have a table that identifies the opportunities to engage students in developing these mathematical concepts in grade K.

COUNTING AND CARDINALITY FOR GRADE K

Grade K

Standard	Materials and Motion Module
Count to tell the number of objects.	
4. Understand the relationship between numbers and quantities; connect counting to cardinality. a. When counting objects, say the number names in the standard order, pairing each object with one and only one number name and each number name with one and only one object. b. Understand that the last number name said tells the number of objects counted. The number of objects is the same regardless of their arrangement or the order in which they were counted. c. Understand that each successive number name refers to a quantity that is one larger.	Inv 1, Part 3, Step 8, Ask questions to guide discussion Inv 1, Part 3 Step 18, Discuss bar graphs Inv 2, Part 2, Step 10, Count number of folds
5. Count to answer "how many?" questions about as many as 20 things arranged in a line, a rectangular array, or a circle, or as many as 10 things in a scattered configuration; given a number from 1–20, count out that many objects.	Inv 3, Part 4, Step 5, Ask questions about the graph Inv 3, Math Extension, Count seams, count pockets and make graphs

Common Core State Standards for Mathematics (National Governors Association Center for Best Practices and Council of Chief State School Officers, 2010).

Trees and Weather Module	Animals Two by Two Module
Inv 1 Part 1, Step 5, Go outdoors	Inv 1 Part 1, Step 4, Guide observations and discussions Inv 2, Part 2, Step 5, Observe shells
	Inv 1, Math Extension, Count the fish in the tanks

OPERATIONS AND ALGEBRAIC THINKING FOR GRADE K

Grade K

Standard	Materials and Motion Module
Understand addition as putting together and adding to, and understand subtraction as taking apart and taking from.	
1. Represent addition and subtraction with objects, fingers, mental images, drawings, sounds (e.g., claps), acting out situations, verbal explanations, expressions, or equations.	
2. Solve addition and subtraction word problems, and add and subtract within 10, e.g., by using objects or drawings to represent the problem.	Inv 3, Math Extension, Make colorful caps

Trees and Weather Module	Animals Two by Two Module
	Inv 1, Math Extension, Add and subtract with fish Inv 2, Math Extension, Use shells for addition and subtraction
	Inv 1, Math Extension, Add and subtract with fish Inv 2, Math Extension, Use shells for addition and subtraction

MEASUREMENT AND DATA FOR GRADE K

Grade K

Standard	Materials and Motion Module
Describe and compare measurable attributes.	
2. Directly compare two objects with a measurable attribute in common, to see which object has "more of"/"less of" the attribute, and describe the difference. *For example, directly compare the heights of two children and describe one child as taller/shorter.*	Inv 4, Math Extension, Math Problem A
Classify objects and count the number of objects in each category.	
3. Classify objects into given categories; count the numbers of objects in each category and sort the categories by count.	Inv 1, Math Extension, List wooden items from home Inv 3, Part 4, Step 5, Ask questions about the graph Inv 3, Part 4, Step 7, Make other graphs

Trees and Weather Module	Animals Two by Two Module
Inv 3, Math Extension, Hang up trunk circumference strings and Measure the circumference strings Inv 2, Part 1, Step 9, Find leaves that go together Inv 2, Part 3, Step 5, Distribute reference cards	Inv 1 Part 1, Step 4, Guide observations and discussions Inv 1 Part 4, Step 4, Guide observations and comparisons Inv 1 Part 5, Step 5, Go outdoors Inv 2, Part 2, Step 7, Seriate shells Inv 3, Part 3, Step 3, Compare worms Inv 3, Math Extension, Compare the lengths of night crawlers
Inv 1, Math Extension, Make a leaf-shape bar graph Inv 4, Part 1, Step 3, Sort tree items	Inv 1, Math Extension, Count the fish in the tanks

GEOMETRY FOR GRADE K

	Standard	Materials and Motion Module
Grade K	**Identify and describe shapes (squares, circles, triangles, rectangles, hexagons, cubes, cones, cylinders, and spheres).**	
	1. Describe objects in the environment using names of shapes, and describe the relative positions of these objects using terms such as *above*, *below*, *beside*, *in front of*, *behind*, and *next to*.	

Trees and Weather Module	Animals Two by Two Module
Inv 2, Part 2, Step 4, Introduce geometric shapes	

Taking FOSS Outdoors

Taking FOSS Outdoors

If we want children to flourish, to become truly empowered, then let us allow them to love the earth before we ask them to save it.

David Sobel, *Beyond Ecophobia*

INTRODUCTION

During its first 20 years, FOSS focused on classroom science. The goal was to develop a scientifically literate population with an ever-growing knowledge of the natural world and the interactions and organizational models that govern and explain it. In recent years, it has become clear that we have a larger responsibility to the students we touch with our program. We have to extend classroom learning into the field to bring the science concepts and principles to life. In the process of validating classroom learning among the schoolyard trees and shrubs, down in the weeds on the asphalt, and in the sky overhead, students will develop a relationship with nature. It is our relationship with natural systems that allows us to care deeply for these systems. In order for students in our schools today to save Earth, and save it they must, they first have to feel the pulse, smell the breath, and hear the music of nature. So pack up your explorer's kit, throw open the door, and join us. We're taking FOSS outdoors.

Contents

WHAT DOES FOSS LOOK LIKE OUTDOORS?

Visualize taking FOSS outdoors: Students exit the classroom in an orderly fashion, their direction and purpose undeterred by the joyful sounds of other students at recess. With focused enthusiasm, the band of young scientists moves toward the edge of the schoolyard. Each student is carrying something, maybe a clipboard for recording, a container for collecting, or a hand lens for observing. Students reach their destination and quickly form a sharing circle. After a brief orientation, students disperse and begin searching the tall grass along the chain-link fence. All are independently recording in their science notebooks, and all are on task. The teacher moves about with intention, speaking to a few students at a time. After several more minutes of this work, the teacher rings a chime. Students freeze, raise one arm, and look at her. She rings the chime again. Students leave their materials in their spots and re-form their sharing circle with their teacher for discussion or additional instructions.

This scenario could be anywhere in the country with a regular classroom teacher using any of the FOSS modules. Taking FOSS outdoors is a natural extension of the classroom work. It looks and feels a lot like standard FOSS activities. Many of the routines you use inside the classroom can be implemented outdoors as well. Success, however, does depend on a few specialized skills and specific preparation to maximize outdoor teaching efficiency.

Expect the enthusiasm, participation, engagement, group discussions, and effort on notebook entries to be heightened during and after an outdoor experience. Even the simplest outdoor activities create a surge of positive energy. It is difficult to determine whether the enthusiasm and commitment students exhibit when doing FOSS outdoors comes from exercising content they already know or from just being outdoors. Students can be a bit louder and more excited when they are learning outdoors, and the space allows for this expansion in energy level, which benefits some students immensely.

GOALS AND OBJECTIVES

The three program goals set down 20 years ago still serve FOSS well. They are: (1) scientific literacy for all students, (2) instructional efficiency and support for teachers, and (3) systemic reform.

The march into the schoolyard has three objectives that relate to the goal for students. First, the outdoor activities **continue and extend the learning** that starts in the classroom. The outdoor activities provide more experience with the content and additional opportunities to practice skills and techniques developed in the classroom.

Second, venturing out provides opportunities for students to discover **applications and examples of classroom content and concepts**. The classroom activities work well for developing sound conceptual science knowledge. That knowledge, however, is constrained by the context in which the concepts are taught. For students to take the next level of ownership of that knowledge, they need to see how it applies and generalizes in the broader context of the world. Leaving the classroom context with a head full of new ideas and new tools for observation enriches the learning.

The third objective is to **connect students with nature**. On the boundaries of the planned, structured experiences are the intangibles that may spark a new relationship with natural systems. It may start with a multisensory experience in the native environment—wind, cold, sunshine, plants, insects, on and on—and advance to an awareness of the diversity of resources surrounding the school. It might evolve into a consciousness of place, followed by a flood of questions about the structure, organization, and operation of the schoolyard ecosystem. When students bond with nature, they have accepted a precious gift, and we have accomplished something important.

MANAGING SPACE

FOSS outdoor activities are designed to be successful in a diversity of schoolyards. Some schoolyards are covered in asphalt, while others have been turned into well-designed outdoor learning environments. Some include large, grassy areas without trees, and others are covered with mulch. One outdoor space may be circled by a variety of mature trees; the next may have recently planted maples and pines scattered about. The space may reflect thoughtful attention or neglect. Nevertheless, FOSS believes that bringing students into the fresh air under a changing sky, into the available outdoor space, will awaken their well-being and stimulate their understanding of science concepts.

Choosing Outdoor Spaces

Whether your school's landscape is wild, manicured, or asphalt, there are more options for outdoor learning spaces than might initially meet the eye. This section will help you choose the best spaces near your school for the FOSS outdoor activity.

Before choosing your outdoor study areas, get to know your outdoor spaces. Look closely at all areas surrounding the school building—even places that students do not normally go. Many seemingly uninteresting monoculture fields are flourishing with a diversity of different grass species and other small flowering plants. Consider the pile of leaves that blew into a corner of your schoolyard; a crack in the concrete; or the ragged, weedy edge of the field where the lawn mower doesn't reach. These are places that provide small animals with what they need—food, shelter, water, and space. Transition zones where vegetation changes from shrubbery to lawn or garden to field can present interesting study sites. As you ponder the learning possibilities in and around your schoolyard, consider these characteristics.

Accessibility. You should be able to walk from your classroom to your outdoor site in 2–5 minutes. Sites farther than 10 minutes away can be considered for special outings, but are not realistic for frequent access. Check out physical access if you have students whose mobility requires consideration. Be aware of slippery surfaces from water or ice, and caution students to be careful.

Purpose. Determine the space needs of the activity. Some activities will require open space, such as a field or blacktop. Other activities work better if students have a more diverse landscape with varying environmental conditions (such as temperature, light intensity, wind). Some activities require a variety of human-made materials to measure or test for certain properties (such as magnets to test for magnetism). Different areas will serve different needs.

Size. The space should be large enough for the class to work comfortably but small enough for you to supervise all students easily. You always need to be able to see all your students, and your students need to be close enough to hear you and your attention signal.

Boundaries. For any space you intend to use, make sure you have clearly defined the boundaries before heading outdoors with your students. Ideally, the landscape will be helpful. For example, stay between the sidewalk and the tree line. If natural markers are not present, you may need to bring along traffic cones or their equivalent to define limits. In general, consider if there are any hazards, such as dangerous debris, poisonous plants, or traffic.

Fostering and Maintaining Diversity

For life science and earth science studies, ideally you want your site to have a variety of living and dead plant matter and a range of environmental conditions. Survey your site to see if it includes places that have been left unmanaged. Even a small wild zone along a fence or behind the maintenance area or an adjacent field can be a valuable resource. It is important for students to see that living things carry on, even in the city, without human assistance.

Enhancing your schoolyard. You may be able to secure a small section of the schoolyard from the school custodian, allowing it to grow wild to compare to the managed school grounds. Consult with your custodian and principal to see if this can be arranged.

Another way to enhance biodiversity is to encourage decomposition by letting fallen leaves and/or lawn clippings to remain on an area of soil over the winter. This gives worms and other decomposers something to eat, which, in turn, provides food for everything else. Make sure all necessary parties are aware of your intentions to leave an area untouched. If you find that you need administrative permission, consider ways to contain and mark the unkempt (but not unloved) area so that it clearly represents an intentional project.

Tread lightly. Your schoolyard study areas will potentially experience some user impact. It is important to teach students to minimize their environmental footprint. Otherwise, the living things they disturb might seek a safer place to live. Unless the class is intentionally collecting specimens, nothing natural should be picked up or removed from the area. This is a good opportunity to introduce the "leave no trace" philosophy, which, in an effort to preserve an area for recreation, encourages us to leave natural objects as we find them.

At some schools, the outdoor space is used by so many classrooms that a system is needed to schedule outdoor activities. A sign-up sheet can be used to reserve outdoor spaces just as is done to reserve other school resources. Check the site the morning before taking the class outdoors to make sure the area is ready for students to investigate.

Weather

Weather can present great challenges and exceptional experiences. Inclement weather can provide an excellent opportunity to study environmental concepts: water drainage, wind impact, plant and animal survival adaptations. (There is nothing like being out in a snowstorm to appreciate the value of insulation!) Making extra preparations to study out in the elements has value. If the activity can be undertaken with some assurance of success, try to make it work. Over time, students acclimate to all sorts of weather and will actually look forward to the challenge of going out in difficult weather.

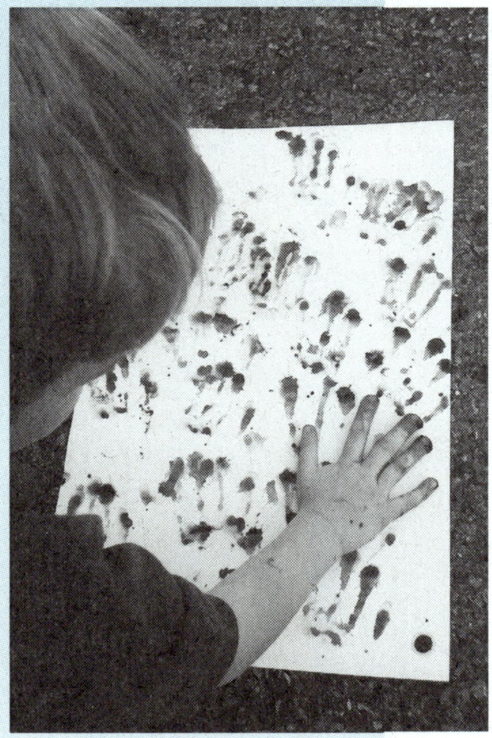

Clothing. The right gear at the right time can make all the difference. Baseball caps stored at school can work well in a light rain and are often essential as sun protection in warmer climates. Baseball caps in a light rain are especially helpful for students who wear glasses. If possible, invest in a set of rain ponchos to make it possible to go out in wet weather. Large trash bags can make very effective, low-cost ponchos. Of course, you will want to model this elegant attire. Communicate regularly with students and family members about upcoming outdoor experiences.

Wind. A stiff breeze can fling your materials into disarray or send notebooks flying. If you anticipate wind, discuss ways to keep materials from blowing away (such as using natural paperweights or taping down nonliving specimens). If there is a protected area where you and your class can take shelter briefly, the activity can continue. You may have to chase down a couple of notebook sheets before students become accustomed to securing papers and other light materials.

Safety and Comfort

Be prepared for the unexpected. Insect stings (ants, bees, wasps, mosquitoes) can be alarmingly painful for young students, particularly if they have not been stung before. Although extremely unlikely in a schoolyard, have a plan developed with students in advance as to how to retreat with purpose if someone accidentally disturbs a nest. You should already know who is allergic and who has never been stung before.

Skin-irritating plants (poison oak, poison ivy, poison sumac, nettles) can certainly put a damper on a field trip. Take a moment, and get to know your local irritants and toxic plants. The rule "leaves of three, let it be" works only for poison ivy and poison oak. Poison sumac has 7–13 leaves on a branch. Stinging nettle feels much like being stung by a jellyfish and can be very frightening for students who have never experienced it. Often, the irritation subsides within a few minutes; do not treat rashes with bleach or rubbing alcohol.

Lyme disease is a treatable bacterial infection, carried by deer ticks. It is present throughout the country, but is particularly present in eastern states. It is possible to get sick without finding a tick bite. If you or your students experience flulike symptoms that are severe enough to see a doctor, make sure that doctor is aware of any outdoor exposure.

If you are out and about in tick country, tuck pant legs into socks and take a few minutes at the end of the trip to pair up and look for obvious ticks on clothing and on the neck and shoulders of a partner.

MANAGING TIME

When to Teach

When you start a new module, anticipate when you might want to go outdoors, and schedule the time. The At a Glance chart in each investigation can help with this planning.

Time of year. If possible, plan the time of year when you will teach particular modules. In the northern tier, life science and earth science modules would be best in the fall or spring. In the southern tier, it might be best to teach life science modules in the winter when it is not uncomfortably hot during the day. Good times to coordinate your outdoor activities with the school calendar include minimum days or other disruptions to the regular schedule, days just before or after school vacations, and days following district testing.

Time of day. Consider the time of day you teach your activities. Established schedules are often difficult to alter, but you might find it advantageous to do so. If you do a lot of seat work in the mornings, you may want to break the routine occasionally with an outdoor activity. Students will return to the classroom refreshed and ready to focus on the next seated activity you have planned.

If you live in a climate where it gets really hot during the school day, you might want to teach outdoors early in the day. Conversely, if you live in a cold climate, you might want to do your winter outdoor activities midday. If you're looking for wildlife (birds, insects, mammals), the best time to go outdoors might be in the morning.

If you plan to use a part of the schoolyard that is heavily populated at predictable times during the day (lunch, physical education), plan to venture out at a time when other activities are minimal.

Stay flexible. If you are studying the **Water and Climate Module**, for example, be prepared to dash out if it rains or snows. One of the delights of outdoor education is going out when nature is putting on a show. Inquiring minds rush out for the experience when timid observers retreat.

Specific times. Some activities require a sunny day. Measuring shadows, solar water heaters, and solar cell investigations require sunshine. It can be tricky to move on without completing specific observations or experiments. Be creative. You may need to proceed with the module and return to the sunny-day activity when the Sun finally comes out.

Instructional Time

An outdoor activity might require 15 minutes, or it might require an hour. Only part of the time budgeted for outdoor learning is actually spent interacting with the schoolyard terrain, plants, and animals. The rest is management.

Travel time. It will take perhaps 10 minutes from the announcement that it is time to decamp for the schoolyard and the time you arrive there. It will take several minutes to describe and distribute materials, get the appropriate clothing, line up, and travel in an orderly fashion to the designated location. Travel back to the classroom will take another 3–4 minutes.

Instructions. Outdoors, students form a sharing circle. It will take 2–4 minutes to review rules, set the boundaries for the activity, describe the challenge, and distribute materials.

Investigation time. Students break into pairs or groups to engage in the outdoor investigation. This might be as short as 8–10 minutes or as long as 30–40 minutes.

Wrap-up. Students return to the sharing circle to share and discuss their discoveries for several minutes.

Classroom follow-up. Frequently, students bring artifacts back to class to display in a classroom museum or to set up for further observation.

Some outdoor activities call for more flexible allocations of time. An activity may call for setup early in the day with periodic monitoring or measuring throughout the day.

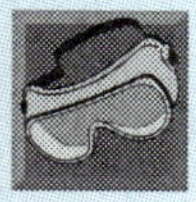

SAFETY NOTE
Students should not disturb or collect live organisms in their natural habitats.

MANAGING MATERIALS

When students step onto the schoolyard, they are field scientists. In the field, there is no lab bench where investigations can be set up, and there is no ready supply of materials. The field equipment must be minimal, portable, and durable so that it can be easily and safely transported from the classroom and back.

Field Equipment

A student's outdoor bag will contain the specific materials needed for the activity of the day as well as some core necessities, such as a hand lens and a writing tool.

Student outdoor bags might contain these basics.

- Pencils/pens
- Hand lens on brightly colored string or yarn
- Colored pencils or crayons
- Measuring tape
- Vials with caps
- Clipboard or notebook
- Seat pad

Note that pens and pencils each have drawbacks: pencil points break, and pen ink freezes in extremely cold weather. Seat pads can simply be several sheets of newspaper covered with a plastic bag.

Your basic teacher's outdoor equipment bag will include a few backup student materials and some items for helping with management.

- Extra pencils, pens, hand lenses, vials, and cups
- Attention signal (chime, whistle, or cowbell)
- Tissues and paper towels
- Basic first-aid kit (adhesive bandages)
- Phone (if leaving the school grounds)
- Student class list (particularly if you teach more than one class) with appropriate student health information and/or permission slips if away from school.

Transporting Materials

Getting materials to and from the outdoor site is a shared responsibility. Students will carry their personal equipment, and class materials can be distributed among students or tackled as a teacher task. Students always carry something to the outdoor site, even when it would be easier for you to carry everything. This reminds students that they are heading out for science, not recess. A hand lens serves as such a token.

Some teachers prefer to have students carry only their clipboards or notebooks and pencils, while the teachers carry all the field equipment in a canvas shopping bag or milk crate to the outdoor home base. Other teachers use a wagon or wheelie crate to transport the equipment. After teaching a few outdoor activities, you will discover what works best for you. Students will get excited when they see you preparing your transport system for an outdoor activity.

Water. Water is often used during outdoor activities. If you are lucky, there will be a tap near your study site. More likely, you will carry water from the school building. Recycled plastic jugs with screw caps and smaller bottles with screw caps are good vessels.

At times, you will want open containers of water, such as buckets or basins (for washing rocks, cleaning containers, and so on). Half-filled buckets can be carried a short distance, but basins should be carried empty and filled from jugs.

You rarely have to bring water back inside. Recycle leftover water by watering schoolyard plants. Make this practice overt to help students develop respect for this vital natural resource.

Creating Outdoor Tools

A sturdy writing surface is essential for science in the schoolyard. A bound notebook (composition book) is excellent. A serviceable clipboard can be made from a piece of cardboard and a binder clip. Use a paper cutter to cut sturdy cardboard slightly larger than a sheet of notebook paper. Place a medium-size binder clip at the top and a large elastic band around the bottom (to keep the paper from flapping up). Tie a pencil on a brightly colored string to the binder clip.

A group writing surface is important sometimes. You can use blue painter's tape to attach a sheet of chart paper temporarily to a wall or clip it onto a chain-link fence with binder clips or clothespins. On windy days, attach all four corners.

A small pack can serve as a hands-free means for students to tote their equipment. Little backpacks are excellent, but a serviceable low-cost satchel can be crafted from a large plastic bag and string. Purchase enough gallon-size zip bags for your class. Punch two holes just *under* the ends of the zipper. (This reduces tearing.) Cut the strings about 1 meter long. Tie sturdy knots that will not come undone. Store the string inside the bag after use to prevent tangling with other bags.

Hand lenses may disappear when students place them on the ground to perform a task. Run bright-colored string or yarn through holes in the lenses for students to wear. If your hand lenses do not already have holes, see if you can get holes drilled through the handles.

MANAGING STUDENTS

Going outdoors regularly is the best way to develop a productive and joyful working relationship with students in the outdoors. When students realize that going outdoors to learn is not a special event but, rather, a science event that will occur routinely, you may be surprised at how quickly they adapt to their expanded, enriched classroom.

Before the First Outing

It is always important to let the school administration know that you and your students will periodically be out of the classroom. If you are planning to leave the school grounds, remember to file a flight plan describing your itinerary, anticipated time of return, and contact partners.

At the beginning of the school year, send a letter home to families, letting them know that learning will extend to the schoolyard and, possibly, beyond. It may be possible to have a signed permission slip for impromptu walking field trips outside the schoolyard. Have families put their contact information and specific student health information on the permission slip. Photocopy these, and have one set of copies in the office and another set in a zip bag in your outdoor equipment bag for emergencies.

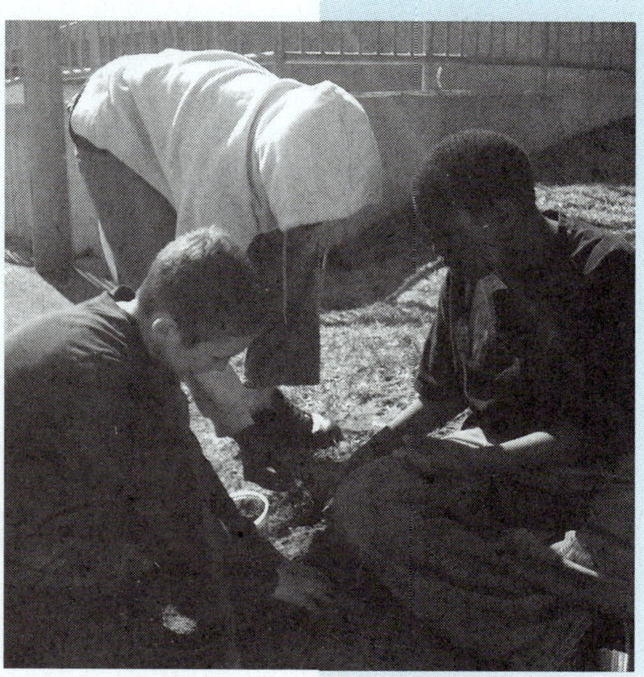

Tell students at the beginning of the year that they will be going outdoors often during science class. Remember to let them know a day in advance that they will be going outdoors. Let them know what it means to dress appropriately. This is especially important in the cold or stormy season when students will need proper clothing for safety and comfort. Your class can go out in any weather if students are dressed appropriately. A consistent system of reminders and clothing preparation will train students to be prepared.

Safety rules. Creating consistent, considerate rules of engagement is important. Learning is enhanced, and behavior problems are largely averted by routines that students participate in and understand.

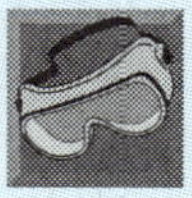

Have a discussion about what students think constitutes proper preparation and behavior for leaving the classroom to study outdoors. (This discussion may be most productive *after* an initial orientation

excursion to survey the schoolyard resources.) Have students generate a list of behaviors that they can adopt and respect. You may want to generate a second list of behaviors that you will agree to as leader of the adventure. Introduce as much formality into the process as you deem important. Develop the idea of a contract that all members of the class sign. Post the contract in your classroom and reference the *Outdoor Safety* poster. Here are the behaviors and rules that should appear on the list.

- Walk quickly and quietly outdoors.
- Outdoor science is not recess.
- Listen to the teacher's instructions.
- Freeze when the teacher rings the bell.
- Stay inside the boundaries.
- Don't make noise near the classrooms.
- Don't injure plants and animals in any way.
- Leave the outdoor environment the way you found it. Never release living organisms into the local environment unless they were collected there.

First Outing

Your first trip to the schoolyard may be a bit chaotic. Students may be distracted by other activities going on, and they may lapse into recess mode. A few precautions will minimize disruptive behaviors.

The path of egress. Determine which doors are available to access your outdoor site. Make sure you follow your school's policy for using auxiliary doors during the school day. Do they need to be closed at all times, or can you prop one open? Are they locked from the outside? You may be able to access keys in order to reduce travel time. When possible, avoid using the door you would normally use for recess. Students have a different mind-set when they walk onto the schoolyard through the "outdoor classroom" door.

Orientation activity. Consider an orientation activity for your first outing. The stated goal might be a site tour to inventory the resources on hand. Your primary agenda, however, is to dissipate the energy generated by the novelty of leaving the classroom during class time. Focus on preparing your transition to the outdoors, moving out in a purposeful and orderly fashion and arriving at your predetermined "home base," a destination that you will always go to initially when you leave the classroom. Form a sharing circle, a process you will use time and again. Tour the schoolyard, proceeding as a whole group, then ask students to walk as individuals for a few minutes, then with a partner for a minute, and finally with a group of four. This gives students a brief experience with each of the four ways they will be organized for various outdoor activities. End with another sharing circle and an orderly return to the classroom.

Challenging students. Sometimes the class will be inattentive and unresponsive. At such times, it is appropriate to direct students back to the classroom. Breaking the contract has consequences, and students need to understand that the opportunity to learn in nature is a privilege. They will remember that day.

In rare instances, you may have an individual student who is regularly not able to comply. Interestingly, this is probably not the student who you anticipated would have difficulty. Often, students who have difficulty with attention and performance in traditional classroom seat work shine and take leadership outdoors. In the case of the noncompliant student, it may be necessary to ask him or her to take a time-out if he or she is able to sit without disrupting others' experiences. If the bad behavior persists, you may be obliged to return to the classroom early. Even so, it is important to give the student a chance to redeem himself or herself the next time you go outdoors.

Routines

Routines are good for management. They impose a measure of self-monitoring because they represent behaviors that are already known and have been practiced. If one person transgresses during a routine, other students are able to intervene to help you with student management. Here are a few routines that may work for you.

Science door. Have you ever watched a group of students pass through the exterior door on their way to recess? As soon as one foot hits the asphalt, they start running and cheering. It is a beautiful sight. Clearly, this is not how you want students to leave the building as you head out for science. One subtle but effective way to distinguish science from recess is to use a different door for science than for recess. Refer to this exit as the "learning door."

Transition behavior. Be explicit about how you want students to walk through the hallways and into the schoolyard. If students exit wildly, simply ring the bell, have them line up inside, and try it again. If this continues to be a problem, return to the classroom and try another day. By consistently showing students that this behavior limits their time outdoors, they will follow your directions.

Home base. Establish a destination in the schoolyard where every outdoor activity will begin. Students should walk directly there after leaving the school building. Choose a place that is level and, if possible, away from classroom windows and popular recess areas.

Sharing circle. When students arrive at home base, they should form a large sharing circle—everyone in a single ring with no students hanging back. This is an effective way to maintain eye contact with all students while you give instructions or share findings. Take a position in the circle where you are facing the Sun. This way, you will know that students won't be distracted by having the Sun's glare in their eyes. A sharing circle is also used to transition from one task to another, to summarize an activity, or anytime you need to regroup.

Techniques for forming a circle vary. One method is "magnetic feet." Students spread their legs to meet their neighbors' feet. Magically, these magnets turn off when you direct them to do so. Students may also stand with hands on hips, elbows touching with neighbors'. Pick or create a method that works with your students.

To speed up the formation of a sharing circle, try the tried-and-true countdown from five, with the objective that everyone is in a proper circle by zero.

Attention signal. Adopt a uniform signal for attention. It is essential that students respond to the attention signal immediately. You may choose the same method you use in the classroom or, if this is not appropriate for the outdoors, try one of these.

- A chime, bell, whistle, or other singular and loud sound. These are appropriate outdoors. When students hear it, they stop, look, and listen.
- Count down from five and when you get to one, students are silent with their hands up. This might not be appropriate outdoors. A countdown from ten can be used to call students back to a sharing circle.
- Clap call and response. You clap a pattern, and students return it by repeating the clapping pattern. This works if students are all nearby.

Focus question. Inquiry-based activities are guided by a question. This pedagogical routine should extend into the schoolyard, too. *Students need to know why they are engaged in the outdoor investigation.* They should

expect to write the focus question in their notebooks at the outset of the investigation and produce an answer at the end of the investigation.

Boundaries. Setting boundaries allows students freedom within a defined space. Because different activities may require different locations, it is always important to be explicit about where students are allowed to travel during the outdoor activity.

Buddy system. You may want to institute a buddy system, particularly if you leave the schoolyard. When participants are paired off, tell them that each individual is responsible at all times for the whereabouts and safety of his or her buddy. It is helpful and fun to number the pairs in order to count off quickly and account for everyone.

Considerations for Students with Disabilities

FOSS evolved from pioneering work done in the 1970s with students with physical disabilities. The legacy of that work is that FOSS investigations incorporate multisensory methods, not only to accommodate students with physical and learning disabilities, but also to maximize information gathering for all students. Strategies that provide opportunities to learn for students with disabilities turn out to be good strategies for all students.

All students benefit from opportunities to experience the natural world outdoors. For students with disabilities, consider how to make the schoolyard accessible and safe so that they can work with a degree of independence. This requires advance planning to make sure that the student, his or her family, the special education teacher, and others involved in the child's school experience are informed and have input into the process.

Whenever a student with a disability is successful in a full-inclusion classroom, there is a behind-the-scenes collaborative effort of caring educators who work together to support the student with just the right amount of scaffolding. In advance of teaching the first outdoor activity, contact the special education teachers in charge of each student's Individualized Education Plan (IEP), and have them review the planned outdoor experiences. Ask the teachers to recommend modifications that will better accommodate and support each student. Invite the special educators to join the class for the outdoor activity.

Attention and language-based disabilities. Students with attention and behavioral issues often thrive when they are engaged in science outdoors. A fenced area will help you to both keep track of students and provide a sense of safety. Having students work with partners (buddies) allows students to look after each other. Provide short, structured opportunities for students to participate in outdoor activities in a clearly defined space, and expand the boundaries and time expectations as students earn your trust and confidence. Many educators have found great success with treating the outdoor activity as a reward for excellent behavior indoors.

Consider students' communication requirements, and plan to bring specialty devices outdoors with you. This might be as simple as some picture cue cards to help enhance your message or an electronic communication device such as a computer.

Physical disabilities and visual impairments. In the Getting Ready section of each outdoor activity, we ask that the teacher decide where the outdoor activity should be taught. You may find that certain locations are better than others for the purposes of developing science concepts and meeting the physical needs of students with disabilities. Get to know your schoolyard really well, and try to experience it as your students do.

One side of most schoolyards is typically a parking lot, and the other three sides have spaces accessible to students for work and play. If you have a student with a physical disability, you need to consider if the terrain provides for good mobility for the student. Often, schoolyards are handicapped accessible because they are covered in asphalt. For many of the FOSS outdoor activities, an asphalt area is appropriate to use. When you want to use a greener location, make sure wheelchairs or crutches will work on these new surfaces. If the surface will be a mobility challenge, see if a paraprofessional or an educational assistant is able to help the student. If someone is not able to join you, consider if a classmate can help. If this is not an option, then consider working at a transition zone where the grass meets the asphalt.

A student with a visual impairment should make a scouting trip to the outdoor site with a mobility instructor to get the lay of the land and to learn where things are located. If the student becomes familiar with and knows how to navigate his or her outdoor surroundings, it will allow for more independence. Even so, during the actual outdoor activity, the student may need someone to quietly describe the terrain ahead and may need a fellow student's arm for balance and security.

If a student struggles with gross motor coordination, uneven ground may present a challenge. Just as you would in the classroom, begin by offering more support, and slowly pull back on this assistance as students become more comfortable with their stamina, security, and endurance with regular outdoor activities.

Sensory sensitivity. For a student with tactile-sensitivity issues, make it clear that he or she may, for example, observe as a classmate digs in the soil to collect a sample. Over time, this student may feel better able to participate by using gloves or by washing his or her hands as soon as the digging is complete. Knowing where each student falls on the continuum of a disability will help you decide when to hold back asking a student to fully participate, when to allow him or her to just observe, and when to give a gentle nudge and expect more active participation.

For students with sensory disorders, the outdoors is often a calming space. Consider where the quietest place in the schoolyard is, and use this more often if you have students with sensitivity to noise.

No matter what the disability, educators have found success taking students outdoors. With advance planning, communication with the student and the special education team, and a little extra effort, you, too, can provide a rich, safe outdoor learning experience for all your students.

TEACHING STRATEGIES

In the beginning, you may find that students regularly use descriptive terms such as "icky," "yucky," and "gross." You may have students who say things such as "I cannot get my clothes dirty. My mom will be mad." Many students are fearful of bugs, wooded areas, and even just sitting on the grass. Often, after a few outdoor activities, these fears and excuses fade away. With patience, persistence, and support, students' resistance may be overcome entirely. If you suspect that your students may be reluctant to work outdoors, structure your first few activities to be low-stress activities. The first few times outdoors can be fairly benign activities with students choosing a comfortable place to just sit (or stand), practicing writing outdoors, and doing simple collecting or counting tasks.

Set the tone. Many teaching strategies that are effective in the classroom work outdoors, too. For example, at the sharing circle, instead of instinctively talking louder (because it is noisy outdoors), kneel down and speak in a loud whisper so that students need to focus to hear you. If students are speaking, put up your silent signal, and wait for silence. The educator's voice sets the tone for the activity.

Take a position. In the sharing circle, position yourself where you have the Sun in your face so that students don't need to squint. If possible, place yourself next to those students who might benefit from a silent look or hand on the shoulder to remind them to be silent.

Meet the challenge. Students who struggle with behavior problems often respond well outdoors when given responsibility. Let the active student carry the heavy jug of water or take the position at the front of the line to lead the class outdoors. For many students, this is all it takes to get them off on the right foot for the outdoor activity.

Students who have the greatest difficulty controlling their behavior indoors are often the leaders when it comes to working in an outdoor space. You may find that students who are not as attentive or cannot sit still inside are the most insistent about quieting down so that the class can get outdoors for science.

Get them writing. Primary students (grades K–2) can fill out a chart on a clipboard outdoors. They are also capable of recording observations outdoors in their notebooks if observation is their only task. Most primary students will need to sit down with their clipboards on their laps or on the ground to do this successfully. In the early years, most writing follows an outdoor activity and is done inside on desks and with the classroom's word wall.

Upper-elementary students (grades 3–5) are capable of writing outdoors. Students will benefit from a quick activity about how to place the notebook or clipboard in the crook of their nonwriting arm for support.

Depending on the activity, you might decide to have students attach their notebooks to a clipboard and place the clipboards in a crate for easy transport and storage. This technique is useful when the ground is moist, when the activity is messy, or when students need to use their hands to complete the activity. The recording will happen immediately after the hands-on activity. Be open to the surprise of how much your students are capable of noticing and recording during and after an outdoor activity.

FLOW OF OUTDOOR ACTIVITIES

The natural flow of a FOSS outdoor activity is slightly different from that of a standard FOSS indoor activity. The steps of a typical outdoor activity are listed below. This list may be helpful if you want to teach more than the handful of outdoor activities in the *Investigations Guide*, or if you want to adapt an indoor activity for schoolyard use.

1. **Prepare for the outdoor activity.**
   - Determine the best location to teach the activity.
   - Check the weather forecast.
   - Make sure students will be dressed appropriately.
   - Prepare materials for distribution.
   - Check the site the morning of the activity.
2. **Set the learning objective.**
   - Present the focus question.
   - Discuss procedures.
3. **Go outdoors.**
   - Gather at the predetermined location.
4. **Model or describe the activity.**
   - Organize students.
   - Define boundaries.
   - Introduce/distribute materials.
5. **Monitor the activity.**
   - Check student engagement.
   - Check student recording.
   - Ask questions.
6. **Share the experience.**
   - Form a sharing circle to discuss experiences.
   - Share thinking.
   - Share answers to the focus question.
7. **Return to class**
   - Make connections to the related indoor activity.
   - Display student work and collections.

EXTENDING BEYOND FOSS OUTDOOR ACTIVITIES

Occasionally, you may stumble upon a serendipitous opportunity. A breeze may launch thousands of twirling seeds from a maple tree, a woodpecker may alight on a tree so close that students can observe it drumming for insects, student-made parachutes may be carried by an updraft high into the sky and out of sight. To your delight, you may spy something you have never seen before. It can happen at any time when you are outdoors!

At special moments like these, our job as educators is to signal students to stop and quietly appreciate the suspension of time. Sometimes, words break the wonder. Trust your instincts at magical moments like these. The answers to questions will come eventually. It is not essential to label the event or even understand it. By inviting students to be alive with their feelings in the moment, you give them a gift for a lifetime.

It is not uncommon for educators to experience the powerful effect of the outdoors on student learning. If you find yourself searching for other outdoor learning opportunities, consider the ones below.

Move activities outdoors. Whatever the subject, students will have more room outdoors to be creative with some activities, and you can worry less about water, sand, and gravel spills. You must still consider how to transport materials, where students will sit, how they will return their project to the classroom, and how to clean up the outdoor space and students' hands before returning to the building.

Use the outdoors for extensions. Extending an inside concept to the outdoors is an excellent way to apply new knowledge. For example, in the **Structures of Life Module**, students grow bush beans hydroponically. If the large leaves fascinate students, go outdoors and see how many kinds of leaves you can find in the schoolyard. Do they all have smooth edges and come to a point at their tips? Go on a leaf hunt, group the leaves by their characteristics, and, eventually, have students tape them into their science notebooks.

There is great value in repeating an indoor activity outdoors. If your students are sanding wood samples inside, follow this up with a trip outdoors to find a stick and sand it. Have you been studying sow bugs? Ask students if they think they know where in the schoolyard they might find these bugs. Applying what students have learned in the classroom and putting that knowledge to work outdoors is an effective way to solidify their understanding. It's also an effective way to informally assess whether students understand the concepts, as well as a method for reinforcing the learning.

NOTE
Remember not to release any classroom animals into schoolyard environments.

Find solitude. Use your outdoor space for silent independent work time. Just as in the classroom, the outdoor space can be a workspace with activities going on. At times, the outdoor space is more of a sanctuary for independent observation and notebook writing. It can be a place for special classroom rituals, awakening awareness of the beauty of nature. Sometimes, it can just be a place to be silent for a minute to awaken the senses and refocus students' energy. Some teachers increase this silent minute to 2, 3, or even 5 minutes. Silence is something to be practiced, and for many students and teachers, this can be challenging. This is a special way to end an outdoor experience and will help students transition into the classroom.

Enhance biodiversity. Modify your schoolyard by adding natural materials, such as logs, rocks, or paving stones. These structures can provide safe havens that may attract more living things. These types of shelters can be particularly helpful if you have an environment without natural shelter from the Sun, such as trees and shrubs. Students can also be involved in the design and implementation of these projects.

Schoolyard modification of this kind requires administrative participation and the support of the school custodian. Marking the area with educational signage can further benefit the enhanced site. If your schoolyard habitat needs your intervention to cultivate biodiversity, understand that it can take a couple of years to get established. Areas completely surrounded by blacktop or concrete can become filled with living things if provided with food, shelter, and water.

Attract wildlife. There are many responsible ways to attract wildlife to your class windows with feeders for birds, squirrels, hummingbirds, or butterflies, as well as many great programs for monitoring these animals. See FOSSweb for ideas for additional wildlife observation projects.

Establish long-term studies. The possibilities for long-term studies are endless, ranging from weather monitoring to seasonal population variation. It can be as simple as adopting an observation location and visiting it monthly to monitor various aspects of change over time. See FOSSweb for ideas for long-term projects.

Create gardens. Planting a garden in raised beds or improved soil is an ambitious option for increasing the biodiversity of your schoolyard. Consider carefully, especially with a vegetable garden, the timing of the school year. In most parts of the country, the time when plants require the most support is during summer vacation. Even if you can get a summer program involved, we suggest starting with indigenous plants that bloom or mature in spring and fall and require little maintenance.

ELEMENTARY-LEVEL ENVIRONMENTAL EDUCATION

In the early 1990s, David Sobel noticed something poignant about children's perceptions of the environment. If a child had been introduced to environmental issues at school that were presented in the context of doom-and-gloom scenarios, the child expressed a heightened sense of anxiety and hopelessness, which Sobel calls ecophobia (Sobel 1996). The implications of his finding should raise a cautionary flag. Sobel is not suggesting that we abandon teaching about the environment in our elementary schools. He is proposing a different approach to environmental education that will bring our children into natural, healthy relationships with environmental issues.

Effective early environmental education should focus on local and ultralocal issues. What is happening in our schoolyards? What factors influence the communities of plants and animals in our neighborhoods? How do changing weather conditions affect the populations around our schools? How are our actions affecting the habitats in our schoolyards? What can we do to enhance natural systems at our schools? Elaborate rain forest projects provide little understanding and have negligible impact on students' connections to nature; researching and installing a butterfly garden or keeping an inventory of the birds in the schoolyard can be transformative. The children from Sobel's 1996 study could tell you how many species in the Amazon were going extinct each minute, but were unfamiliar with the most common plants in their schoolyard.

Time outdoors during the school day is beneficial for student learning. Students who are exposed to hands-on experiences in their local environment often become enthusiastic, self-motivated learners and, typically, academically outperform their peers who do not have these learning opportunities (Liebermann and Hoody 1998). Children are able to pay attention for longer periods of time outdoors on the same assignment and are more focused when they return to their indoor class work (Louv 2008).

Research has produced evidence that using the schoolyard is an effective way to enhance student learning. Texas A&M University, in conjunction with the Texas Education Agency, conducted a meta-analysis of the research in order to identify and rank effective instructional methods for science education and to define how best to improve student achievement. The highest-ranked teaching strategy was Enhanced Context Strategies, which included taking meaningful field trips and using the schoolyard for activities (Scott et al. 2005).

Taking FOSS Outdoors

Students' attitudes toward learning are influenced by simple outdoor experiences. In one study (Shaw and Terrance 1981), students who experienced outdoor instruction reported that, in general, they enjoyed school more and felt more supported and trusted by their teacher than they had prior to the outdoor experiences. These pretest/posttest differences were more pronounced for students who had been identified as being "uninvolved" in the classroom activities. Also, this student perception was a lasting effect that carried over to the regular classroom activities weeks later.

Perhaps the most important benefit of incorporating the outdoors into the traditional school learning environment is that it offers opportunities for students to synthesize concepts and personal experience by applying what they have learned to a new environment.

FOSS outdoor activities will help you focus on age-appropriate environmental topics and enable you to create meaningful and personal connections between your students and their local environment. When students can openly explore the environment, they can create meaningful connections to their learning and establish positive relationships with nature. You'll be amazed by what students notice.

Here's the good news. If you focus on inquiry and direct experience instead of problems, it takes remarkably little guidance for students to make positive, empowering, lifelong connections to nature. One insightful young man explained, "My video games have a pattern that is always the same, but nature is like a game that is different every time you play." As an educator, you can draw out that sense of wonder and curiosity for students while simultaneously helping them build a solid science foundation.

REFERENCES

Liebermann, G., and L. Hoody. 1998. *Closing the Achievement Gap: Using the Environment as an Integrated Context for Learning; Results of a National Study.* San Diego: State Education and Environment Roundtable.

Louv, R. 2008. *Last Child in the Woods: Saving Our Children from Nature-Deficit Disorder.* New York: Workman Publishing.

Scott, T. et al. 2005. *Texas Science Initiative Meta-Analysis of National Research Regarding Science Teaching.* Texas Education Agency.

Shaw, T. J., and J. M. Terrance. 1981. "Involved and Uninvolved Student Perceptions in Indoor and Outdoor School Settings." *Journal of Early Adolescence,* 1:135–146.

Sobel, D. 1996. *Beyond Ecophobia: Reclaiming the Heart in Nature Education.* Great Barrington, MA: Orion Society.

ACKNOWLEDGMENTS

The Taking FOSS Outdoors initiative got its start through a collaboration with the Boston Schoolyard Initiative (BSI). In 2004, BSI began developing an approach to teaching science that routinely takes students into the schoolyard to test, apply, and explore core science concepts and skills. As part of this project, BSI developed *Science in the Schoolyard Guides*™ for 12 FOSS modules and a companion *Science in the Schoolyard*™ DVD. In partnership with the City of Boston, BSI designs and builds schoolyards that provide a rich environment for teaching, learning, and play. For more information on BSI, *Science in the Schoolyard*, or BSI's *Outdoor Writer's Workshop*™ professional development program and materials, see www.schoolyards.org.

Teacher Masters

LETTER TO FAMILY

Science News

Dear Family,

Our class is beginning a study of materials and motion. As materials engineers, we will be studying the properties of different kinds of wood, paper, and fabric. We'll investigate how wood and paper can be processed into products. We'll sand wood, make simulated plywood and particleboard, recycle paper, and make papier-mâché bowls. We will study how fabrics are made and discover how they interact with water. We will focus on the reuse and recycling of materials to conserve natural resources. We will be setting up a recycling center in our classroom. You can enrich this experience by having your child participate in the recycling of paper, metal, glass, and plastic at home.

Help us by gathering wood scraps and interesting and colorful paper and fabric discards for making our structure design project. Please send wood, paper, and fabric scraps by ______________ (date).

We will also investigate motion—how to get something to start moving and make it stop. We will change the strength of pushes and pulls on rolling objects to see how that affects speed, and look at how we can modify direction of moving objects to meet goals.

After we do the various activities in class, your child may ask you to help him or her do things at home, such as temporarily label objects with "paper" or "wood"; waterlog a stick; take boxes apart; or make collages, envelopes, drinking cups, or paper hats. You can get more information on this module by going to www.FOSSweb.com.

Sincerely,

FOSS

Science News

Dear Family,

Our class is beginning a study of materials and motion. As materials engineers, we will be studying the properties of different kinds of wood, paper, and fabric. We'll investigate how wood and paper can be processed into products. We'll sand wood, make simulated plywood and particleboard, recycle paper, and make papier-mâché bowls. We will study how fabrics are made and discover how they interact with water. We will focus on the reuse and recycling of materials to conserve natural resources. We will be setting up a recycling center in our classroom. You can enrich this experience by having your child participate in the recycling of paper, metal, glass, and plastic at home.

Help us by gathering wood scraps and interesting and colorful paper and fabric discards for making our structure design project. Please send wood, paper, and fabric scraps by ______________ (date).

We will also investigate motion—how to get something to start moving and make it stop. We will change the strength of pushes and pulls on rolling objects to see how that affects speed, and look at how we can modify direction of moving objects to meet goals.

After we do the various activities in class, your child may ask you to help him or her do things at home, such as temporarily label objects with "paper" or "wood"; waterlog a stick; take boxes apart; or make collages, envelopes, drinking cups, or paper hats. You can get more information on this module by going to www.FOSSweb.com.

Sincerely,

FOSS

FOCUS QUESTIONS A

Inv. 1, Part 1: **Where does wood come from? What is made of wood?**

Inv. 1, Part 2: **What happens when wood gets wet?**

Inv. 1, Part 3: **How can you sink wood? How many passengers can a wood raft hold?**

Inv. 1, Part 4: **How can you change the shape of wood?**

Inv. 1, Part 5: **How are sawdust and shavings the same? How are sawdust and shavings different?**

Inv. 1, Part 6: **How is particleboard made?**

Inv. 1, Part 7: **How is plywood made?**

FOCUS QUESTIONS B

Inv. 2, Part 1: **What is made of paper?**

Inv. 2, Part 2: **What makes paper good for writing?**
What makes paper easy to fold?

Inv. 2, Part 3: **What happens when water gets on paper?**

Inv. 2, Part 4: **How can new paper be made from old paper?**

Inv. 2, Part 5: **How can paper be made strong to form a bowl?**

Inv. 3, Part 1: **How are fabrics different?**
What is made of fabric?

Inv. 3, Part 2: **How is fabric made?**

Inv. 3, Part 3: **What happens when water gets on fabric?**

Inv. 3, Part 4: **How are different kinds of fabric used?**

Inv. 3, Part 5: **How can we conserve natural resources?**

Inv. 3, Part 6: **What happens to water in sunshine and shade?**
How can we design a structure to keep water cool in sunshine?

FOCUS QUESTIONS C

Inv. 4, Part 1: **What causes objects to move?**

Inv. 4, Part 2: **What happens when objects collide?**

Inv.4, Part 3: **Where can balls roll on the schoolyard?**

Inv. 4, Part 4: **How can we change how far a balloon rocket travels?**

WOOD LABELS

This is made of **wood**.	This is made of **wood**.
This is made of **wood**.	This is made of **wood**.
This is made of **wood**.	This is made of **wood**.
This is made of **wood**.	This is made of **wood**.
This is made of **wood**.	This is made of **wood**.
This is made of **wood**.	This is made of **wood**.
This is made of **wood**.	This is made of **wood**.

CENTER INSTRUCTIONS—WOOD AND WATER

Materials

- Droppers
- Small objects that sink
- Water in cups
- Water in clear basins
- Wood samples in containers
- Sponges
- Paper towels
- Newspaper

Set Up the Center

Put a dropper at each student's place and a cup of water for each pair of students to share. Have basins of water nearby so that they can be easily moved to the center when needed. Set out the containers of wood samples for students to take as needed. Each student needs one dry sample of each of the five kinds of wood.

Guide the Investigation

1. **Drop water on cedar and particleboard.** Each student gets one sample of cedar and particleboard. Students put one drop of water on one side of both samples and observe. Then they put a drop of water on the end of both samples.
2. **Monitor progress and discussion.** Discuss what happens when students drop water on the wood samples. Guide observations by asking questions.
   - *Does water sit on the pieces of wood, or does it soak in?*
   - *How does it soak in? (Does it get absorbed right away, or does it take some time?)*
   - *Does the color of the wood change when it gets wet?*
3. **Drop water on the other samples.** Let students continue by dropping water on the three other wood samples (basswood, plywood, and pine).
4. **Use basins of water.** Collect the cups of water and droppers. Ask students what they think will happen if they put the wood samples in a tub of water. Put a basin of water on the table for each pair of students to share. Let students put their samples of wood into the water, one at a time. (All the samples should float.)

   NOTE: Students often discover that wet samples stick together. Challenge students to see how many samples will stick together when they pick them up.
5. **Add to the word wall.** Add new or important words that students use to the word wall.
6. **Clean up.** Remove the water basins from the table, but keep them handy for the next group. After all the groups are finished, spread the wood samples out on newspaper to dry overnight.

Vocabulary

Try to include these words in discussions with students:
absorb, communicate, float, sink, soak, spread

CENTER INSTRUCTIONS—TESTING A RAFT A

Materials

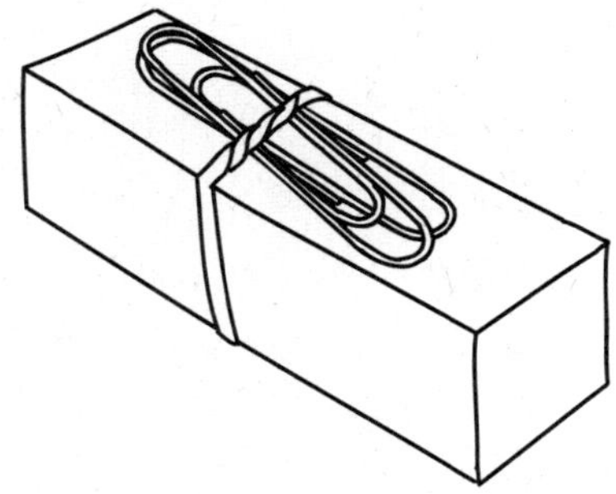

Basins half filled with water
Jumbo paper clips in cups
Rubber bands in a cup
Sponges
Pine samples
Plywood samples
Paper towels
Newspaper

Set Up the Center

Set a clear basin half filled with water on the table for each pair of students. Put cups of jumbo paper clips near each basin and a cup of rubber bands in the center of the table within easy reach of all students. Have pine and plywood samples ready.

Guide the Investigation

1. **Discuss sinking.** Show students the paper clips and rubber bands. Ask students to see if they can use these materials to sink the two kinds of wood.
2. **Begin the exploration.** Distribute one sample of pine and plywood to each student. Let students explore freely. (If students are stumped, show them how to put a rubber band around a piece of wood and slide a paper clip under the rubber band.)
3. **Quantify the results.** When students can attach paper clips to a sample, challenge them to find out how many paper clips are required to sink each piece of wood.

   NOTE: Young students may not use a systematic method to determine how many paper clips it takes to sink the wood. They love to attach as many paper clips as they can. Let them use their own methods, regardless of accuracy.
4. **Ask questions to guide discussion.**
   - *Were you able to sink the wood by attaching paper clips?*
   - *How many paper clips did you use to sink the wood?*
   - *Let's compare the two samples. Does it take the same number of paper clips to sink both kinds of wood?*
   - *Does it make a difference where you put the paper clips on the wood? All on one side? Evenly distributed around all sides?*
5. **Add to the word wall.** Add new or important words that students use to the word wall.
6. **Clean up.** Have students remove the paper clips and rubber bands and return them to their containers for the next group. Lay the wet wood samples on newspaper to dry. After all the groups are finished, spread the paper clips on towels to dry.

Vocabulary

Try to include these words in discussions with students:
above, below, compare, float, sink, test

CENTER INSTRUCTIONS—TESTING A RAFT B

Materials

Particleboard samples
Cedar samples
Sponges
Newspaper
Basins half filled with water
Rubber bands in a cup
Jumbo paper clips in cups
Paper towels

Set Up the Center

Set a basin half filled with water on the table for each pair of students. Put a cup of paper clips near each basin and a cup of rubber bands in the center of the table within easy reach of all students. Have the two kinds of wood ready for distribution.

Guide the Investigation

1. **Sink the wood samples.** Have students use the following procedure to sink each piece of wood in a systematic fashion. Give the wood samples to students one at a time, cedar first, then particleboard. (A sample is not sunk until it lies flat on the bottom of the basin.)
   a. *Put a rubber band around the sample.*
   b. *Slip one paper clip under the rubber band, and place the sample in the water to see if it floats.*
   c. *Slip another paper clip under the rubber band on the sample. Test to see if it still floats.*
   d. *Continue this procedure, testing after each paper clip is added, until the sample sinks.*
   e. *Set the piece of wood aside with the rubber band and paper clips still in place.*
2. **Make bar graphs.** Work with one sample at a time. Remove the paper clips from the rafts. Have students place the sample of wood on the table and lay the paper clips in one of the two ways shown.

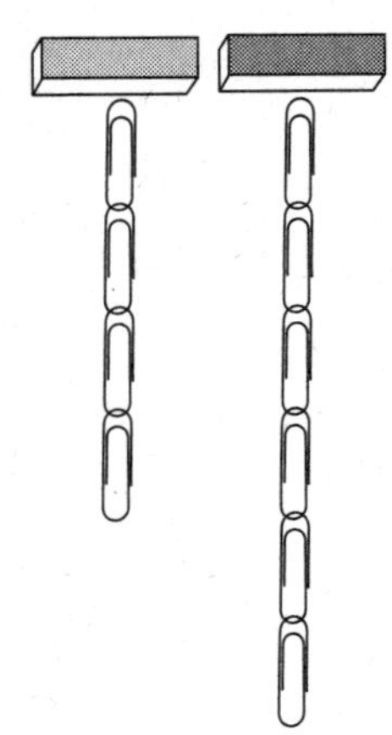

3. **Ask questions about the graph.**
   - *Which piece of wood took more paper clips to sink, the cedar or the particleboard?*
   - *How could you tell without counting the paper clips?*
   - *Which piece of wood took fewer paper clips to sink?*
   - *How many paper clips did it take to sink the cedar? The particleboard?*
4. **Add to the word wall.** Add new or important words that students use to the word wall.

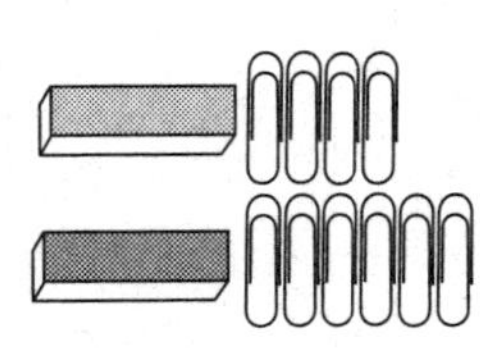

5. **Clean up.** To get ready for the next group, return paper clips and rubber bands to their containers and put wood samples on newspaper to dry overnight. After all the groups are finished, spread the paper clips on paper towels to dry.

Vocabulary

Try to include these words in discussions: *fewer, float, graph, less, more, raft, sink, test*

CENTER INSTRUCTIONS—SANDING WOOD

Materials

Sandpaper
Paper plates
Plastic cup
Basswood samples for sanding
Small sticks

Set Up the Center

Put a paper plate and a piece of basswood at each student's place. Keep the sandpaper out of reach until you are ready to have students begin.

Guide the Investigation

1. **Let the sanding begin.** When students are settled, distribute one piece of sandpaper to each student. Let them begin sanding the basswood. Tell them to keep the dust from the sanding on their paper plates.
2. **Discuss the results.** As students sand the wood, ask them to share their observations about what happens when they rub the wood with the sandpaper. Guide the discussion by asking questions. (Because the wood they are sanding is fairly smooth, focus on how the shape of the wood can be changed.)
   - *How do you change the shape of the wood?*
   - *What is the best technique for sanding?*
   - *How does the wood feel after you sand it?*
   - *What has fallen onto the plate?*
   - *Where does the wood dust come from?*
   - *How does the dust feel?*
   - *What are the black dots mixed in with the sawdust? Where do they come from?*
3. **Collect the sawdust.** (Adult does this.) Use the plastic cup to collect the sawdust from each student's plate. Tell students that you are saving the dust for another project.
4. **Sand a small stick.** Have each student find or select a small stick to sand. Have students sand the stick and compare it to the basswood.
5. **Add to the word wall.** Add new or important words that students use to the word wall.
6. **Clean up.** Prepare the center for the next group. The paper plates should be reused. Check the wood pieces and sandpaper and reuse or replace.

Vocabulary

Try to include these words in discussions with students:
change, sand, sandpaper, sawdust, shape, woodworker

CENTER INSTRUCTIONS—SAWDUST AND SHAVINGS

Materials

Shavings in containers
Sawdust in cups
Screens
Paper plates
Plastic spoon
Paper towels
Water in pitcher
Craft sticks

Set Up the Center

For each student, put one heaping spoonful of sawdust in one plastic cup and one heaping spoonful of wood shavings in a container. Put all the materials on a table or counter close by so they will be within easy reach for distribution to students.

Guide the Investigation

1. **Explore the sawdust and shavings.** Give each student a cup of sawdust and a craft stick. Let students explore the sawdust with the stick for a minute. Give each student a container of wood shavings to explore for a minute.
2. **Combine the two materials.** Ask students to *pour the sawdust into the wood-shavings* container. They should observe what happens, then stir the mixture.
3. **Drop the sawdust and shavings in water.** Half fill each student's empty cup with water. Ask what will happen if students sprinkle the sawdust and shavings on the water. Let them try it. Tell them not to stir and to look through the side of the cup.
4. **Discuss the relative position of the wood in the water.** After 2 minutes, have students stir.
   - *Where are the pieces of wood in the water?*
   - *Are they above the surface, below the surface, on the bottom?*

   Help students describe the location of the shavings in relation to either the surface of the water or the bottom of the cup.
5. **Discuss results.** Explain that when wood sinks, we say it is waterlogged.
6. **Strain out the wood pieces.** Have students place a screen on top of the container and pour the shavings mixture onto the screen. Have them dump the sawdust and shavings caught on the screen onto a paper plate to dry overnight. Ask,
   - *How did the sawdust and wood shavings change after you added water?*
7. **Clean up.** Get the center ready for the next group. Dry the cups and containers. Put one heaping spoonful of sawdust in one plastic cup and one heaping spoonful of wood shavings in a container for each student.

Vocabulary

Try to include these words in discussions with students:
mixture, sawdust, screen, shavings, waterlogged

CENTER INSTRUCTIONS—MAKING PARTICLEBOARD

Materials

Craft sticks
Paper plates
Paper towels
Scratch paper
Plastic spoons
Particleboard samples
Cornstarch matrix in containers
Sawdust/shavings mixture in cups
Plastic cups

Set Up the Center

Put a plastic cup with two heaping spoonfuls of the dry sawdust/shavings mixtures at each student's place. Keep the other materials at the end of the table where you have easy access to them.

Guide the Investigation

1. **Review working with sawdust.** Tell students they will use homemade glue to stick the sawdust together to make particleboard, or sawdust wood.
2. **Distribute the cornstarch matrix.** Show students the glue made from cornstarch. Pass the containers of matrix around. Help each student put one heaping spoonful in his or her cup of sawdust.
3. **Stir up the mixture.** Students use craft sticks to mix the sawdust and matrix. When it is mixed, have them dump it on a paper plate and continue mixing with their fingers. The final product should be the consistency of modeling dough. Add matrix or sawdust if necessary.
4. **Shape the particleboard.** Have students shape the sawdust dough in any way they want. Have students make observations while they are working and compare their projects to other students' projects.
5. **Let the wood dry overnight.** Set the pieces of particleboard on scratch paper to dry overnight. Write students' names on the paper next to the wood pieces.
6. **Add to the word wall.** Add new or important words that students use to the word wall.
7. **Clean up.** Wipe out the used cups with a paper towel. Refill them with two spoonfuls of sawdust. Refill the containers with cornstarch matrix as needed for the next group. Reuse the paper plates.

Vocabulary

Try to include these words in discussions with students:
cornstarch, glue, mix, particleboard

CENTER INSTRUCTIONS—MAKING PLYWOOD

Materials

Newspaper mats
White glue
Paper towels
1/4" plywood samples
Thin wood pieces
Pencil
Craft sticks

Set Up the Center

Put a sheet of folded newspaper at each student's place. Pile the thin wood pieces in the center of the table where students can easily reach them. Have the white glue ready for distribution when needed. Have the 1/4" plywood samples ready for students to compare to what they make.

Guide the Investigation

1. **Make plywood.** Have each student choose three thin wood pieces. Go around the group and squeeze a quarter-sized pool of glue onto the far side of each student's newspaper mat. Let students begin gluing the pieces together.

   NOTE: Make sure students put glue on both surfaces they are gluing together, spread the glue over the entire surface of each piece, and hold the pieces together for a count of 20 after each piece of wood is added.

2. **Add to the word wall.** Add new or important words that students use to the word wall.

3. **Make comparisons.** Put the 1/4" plywood samples around the table. Ask students to compare the plywood samples to the plywood they made.

4. **Clean up.** Find a location where the wood can dry overnight. Label students' pieces of plywood by writing their names on the plywood. Replace the newspaper mats for the next group of students, or use the same sheets turned inside out if they are not covered with too much glue.

Vocabulary

Try to include these words in discussions with students:
break, laminate, layer, plywood, strong

HOME/SCHOOL CONNECTION A

Investigation 1: Getting to Know Wood

At school, we took a close look at five kinds of wood to discover all the ways they were alike and different. Students got to know the wood so well that they went on a hunt in the room, searching for wood samples that matched their own. Along the way, they discovered many things that are made from wood. Here are two ways to practice the vocabulary and observation skills your child is developing.

- Give a clue: "I'm thinking of something that is made of wood, and it is (use descriptive words, such as *round*, *big*, *painted*, *flat*)." Take turns describing and identifying wood in use around your home or in your neighborhood.
- Have your child search for four different ways that wood is used around your home or neighborhood. Have him or her label the objects using the labels below (cut them from the sheet) and draw each object in one of these boxes.

This is made of **wood**.	This is made of **wood**.
This is made of **wood**.	This is made of **wood**.

HOME/SCHOOL CONNECTION B

Investigation 1: Getting to Know Wood

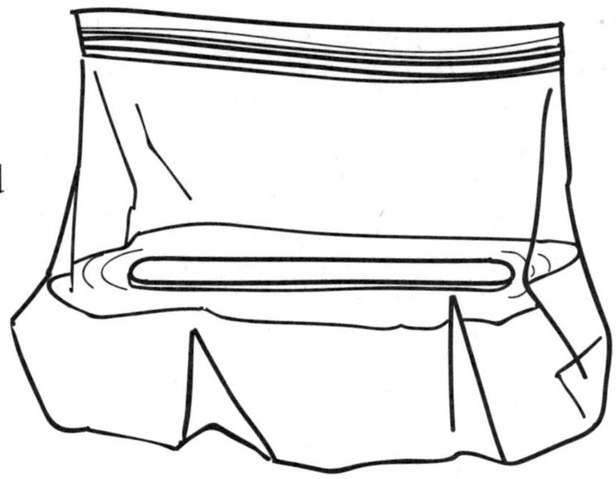

We have been investigating what happens when water and wood come together. We've dropped water on different kinds of wood, floated wood, and sunk wood. One discovery we made was that tiny pieces of wood can become waterlogged. We are wondering if larger pieces of wood will become waterlogged, too. Here is one way to find out.

Materials

- 1 Craft stick
- 1 Plastic zip bag or bowl
- • Water

Investigation

1. Fill a zip bag or bowl about one-third full of water.
2. Float the craft stick in the water. Leave the bowl or bag out where you can see it. If you have a bulletin board, you can tack the zip bag to it.
3. See how long the stick takes to become waterlogged. Ask your child how he or she will know if it is waterlogged. (It will sink to the bottom. This might happen as quickly as overnight or might take a week or more.)

PAPER LABELS

This is made of **paper**.	This is made of **paper**.
This is made of **paper**.	This is made of **paper**.
This is made of **paper**.	This is made of **paper**.
This is made of **paper**.	This is made of **paper**.
This is made of **paper**.	This is made of **paper**.
This is made of **paper**.	This is made of **paper**.
This is made of **paper**.	This is made of **paper**.

CENTER INSTRUCTIONS—USING PAPER A

Materials

Pencils
Crayons
Marking pens
Paper samples in containers

Set Up the Center

Place the containers of paper samples close to the center so you and students can pick up samples as needed. Have pencils, crayons, and marking pens on the table for students to use.

Guide the Investigation

1. **Compare the samples.** Place the containers of tagboard and paper-towel samples on the table. Ask students to take one sample of each paper. Discuss how the papers are alike and how they are different.

2. **Mark the samples with pencils and crayons.** Tell students to make marks on each sample with pencils and then with crayons. Discuss which paper provides the best surface for writing or drawing.

   NOTE: It is OK if the paper rips or gets a hole in it. That's valuable information.

3. **Use a marker.** Tell students to make marks on the paper samples with a marking pen. Compare the results of the marks on the two kinds of paper. Discuss how the ink from the marking pen soaked into the paper towel (or the paper towel absorbed the ink). Compare this to the tagboard, which did not absorb much ink.

   CAUTION: Roll up sleeves! Marking-pen ink on waxed paper can make a mess.

4. **Mark the other samples.** Let students make marks on the other paper samples, using the pencils, crayons, and marking pens. Ask questions to guide students' discoveries.
   - *Is this a good paper for writing a note? Why or why not?*
   - *Does the ink from the marking pen soak into this paper or stay on the top?*
   - *What makes paper easy to write or draw on? What makes it hard to write or draw on?*
   - *Find two papers that feel the same when you write on them. How are they the same?*
   - *Find two papers that feel different when you write on them. How are they different?*
   - *How did you feel when you were writing on the corrugated (bumpy) paper?*

5. **Clean up.** Collect all the paper samples; they will be reused in the folding center.

Vocabulary

Try to include these words in discussions with students:
absorb, bumpy, slick, tear, texture

CENTER INSTRUCTIONS—USING PAPER B

Materials

Paper samples in containers

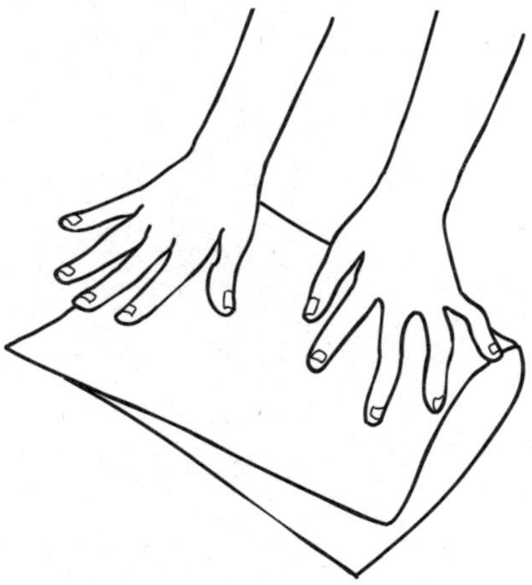

Set Up the Center

Put the newsprint samples at the center of the table. Have other paper samples ready to use as needed.

Guide the Investigation

1. **Fold the newsprint.** Let each student take one newsprint sample. Have students fold it in half once. Monitor them to make sure they understand the concept of folding in half. Have them continue folding until the paper can't be folded any more.

2. **Count the number of folds.** Have students count the number of folds by unfolding their samples and counting each time they open up a fold.

3. **Fold the other papers.** Put the other containers of paper samples on the table. Let students take one of each to see how many times they are able to fold each sample.

4. **Ask questions to guide discussion.**
   - *What kinds of paper are easy to fold, and what kinds are hard to fold?*
   - *What makes some kinds of paper easy to fold and other kinds hard?*
   - *What does a sheet of paper look like after it has been folded and then opened up?*
   - *How many times do you think you could fold a facial tissue?*
   - *Why would people want to fold paper?*
   - *What could you make by folding paper?*

5. **Sort the paper samples.** When students have finished folding all their samples, ask them to sort the samples into two piles: those that were easy to fold and those that were more difficult to fold. Discuss the properties of paper that affect how easy it is to fold a piece of paper.

6. **Add to the word wall.** Add new or important words that students use to the word wall.

7. **Clean up.** Collect the samples and save them for future activities. Each new group will need samples that have not been previously written on or folded.

Vocabulary

Try to include these words in discussions with students:
bend, corner, crease, flat, fold, half, thick, thin

CENTER INSTRUCTIONS—PAPER AND WATER

Materials

Paper samples in containers
Sponges
Droppers
Water in basins
Water in cups

Set Up the Center

Place a dropper and a cup with 1 cm of water at each student's place. Have the containers of paper samples ready for students to pick up at the center. Keep sponges on hand for possible spills.

Guide the Investigation

1. **Drop water on the paper-towel and tagboard samples.** Have students pick one sample of paper towel and tagboard. Ask them to put *one drop at a time* on the two samples and observe. Ask students to compare the two samples.

2. **Drop water on remaining samples.** Let students choose and test two more samples. Continue until students' interest wanes.

3. **Ask questions to guide observations and discussion.**
   - *Compare what the water does on each sample. How is it the same, and how is it different?*
   - *Which samples would be good for soaking up spills? Why do you think so?*
   - *Are there any samples that would be good to wrap things in to protect them from water?*
   - *Which papers stayed the same, and which changed? How did they change?*

4. **Add to the word wall.** Add new or important words that students use to the word wall.

5. **Soak paper samples overnight.** After students have finished dropping water on the paper samples, ask them what they think the samples would look like if they soaked in water overnight. Let students choose one of each sample to put in a basin to soak overnight.

6. **Clean up.** Collect the wet paper samples and spread them out to dry; they can be used again. After the last group, set the droppers and cups on towels to dry.

Vocabulary

Try to include these words in discussions with students:
bead up, change, drop, dropper, submerge, wet

CENTER INSTRUCTIONS—PAPER RECYCLING

Materials

- Newspaper mats
- Bottles with caps
- Toilet tissue
- Water in pitcher
- Waxed paper squares
- Sponges
- Basins
- Containers
- Screens
- Self-stick notes

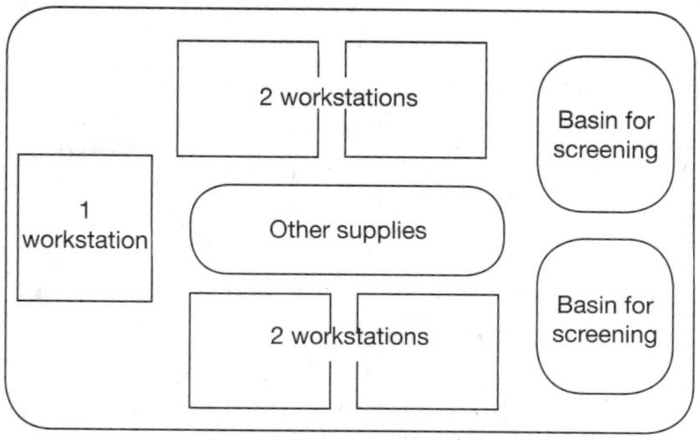

Paper-recycling center

Set Up the Center

At each student's place, put two folded sheets of newspaper, a bottle with a cap, and an eight-sheet length of toilet tissue. Fill a pitcher with water and keep it near the center. Place the waxed-paper pieces and sponges in the center of the table. At one end of the table, set up two screening stations, each with a basin. Put a container in each basin. Keep the screens nearby.

Guide the Investigation

1. **Make the paper pulp.** Have students stuff the toilet tissue into the empty bottles. Add water until the bottles are about two-thirds full. Have students screw the caps securely on the bottles and shake them. It takes about 3–5 minutes for the paper to break down into pulp.
2. **Screen the pulp.** Call two students at a time to the screening station. Have each student follow this procedure.
   a. Place a screen on a container in a basin.
   b. Pour the pulp onto the screen. (Students will need to shake the bottles or bump the end of the bottles to get the pulp out.) Add a little more water if all the pulp won't come out.
   c. Distribute the pulp over the screen.
   d. Place a second screen on top of the pulp to make a screen-and-pulp "sandwich."
3. **Soak up the water and roll out the paper.** Send students back to their workstations with the screen-and-pulp sandwiches to finish the process.
   a. Lay the screen-and-pulp sandwich on a newspaper blotter.
   b. Place a sponge on top of the screen and press to remove as much water as possible.
   c. Take off the top screen. Place a piece of waxed paper on the pulp. Flip the screen (with the pulp) over onto the waxed paper. Then lift off the screen.
   d. Roll a bottle over the paper pulp to press more water out and make it a thin, smooth piece of paper. Use the sponge to blot water from the waxed paper.

   NOTE: Roll from the middle to the edges to prevent the paper from curling around the bottle.
4. **Ask questions to guide discussion.**
   - *What happens to the pieces of tissue when you add the water? When you shake them up?*
   - *Where does the water go when you press the pulp with the sponge?*
   - *What happens when you roll the bottle over the paper?*
   - *What does the paper look like when it begins to dry?*
   - *How does the recycled paper compare to the tissue used to make the pulp?*

Vocabulary

Try to include these words in discussions with students:

absorb, blot, fiber, flip, pulp, recycling, rolling, screen

CENTER INSTRUCTIONS—PAPIER-MÂCHÉ

Materials

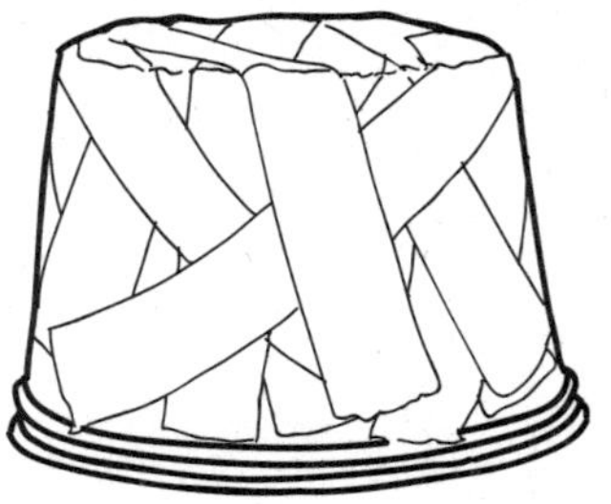

Newspaper, folded, half sheets
Newspaper to cover tables
Containers, 1/2 L
Wheat paste
Plastic cups
Self-stick notes

Set Up the Center

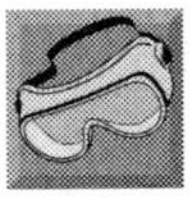

Place a half sheet of folded newspaper and a 1/2 L container at each student's place. Pour some wheat paste in a cup for each student. Make sure no student is allergic to wheat.

Guide the Investigation

1. **Tear the newspaper.** Have students tear their folded sheets of newspaper into strips 2.5 cm (1") wide, then into 10 cm (4") lengths. Have students tear the entire sheet of newspaper.

2. **Begin to papier-mâché.** Have students follow these directions.

   a. Place the container upside down.
   b. Cover the container with a thin layer of wheat paste, spreading it evenly with your fingers.
   c. Lay a paper strip over the pasted bowl, and cover it with a layer of paste. Be sure to use enough paste to completely soak the paper, then wipe the excess off.
   d. Follow the same procedure using a second strip of paper, overlapping the first strip already on the container.
   e. Cover the entire outside of the container with several layers of paper strips, down to but not over the raised area at the rim of the container.

3. **Ask questions to guide discussion.**
   - *What does the wheat paste look like?*
   - *How does the wheat paste feel? How does it feel when it dries on your hands?*
   - *What happens to the paper when it is soaked in paste?*
   - *Is the paper easy or hard to tear when it is soaked in paste?*

4. **Let the papier-mâché dry overnight.** Use self-stick notes to label each student's project on the inside of the container. Set the bowls out to dry.

5. **Add to the word wall.** Add new or important words that students use to the word wall.

6. **Clean up.** Clean the work area for the next students.

Vocabulary

Try to include these words in discussions with students:
across, around, dry, flour, layer, mold, over, papier-mâché, stiff, strip, tear, wheat paste

MATH EXTENSION—PAPER BOXES

Investigation 2: Getting to Know Paper

Materials

Paper boxes
Crayons or marking pens
Transparent tape
Large sheets of newsprint or newspaper

Investigation

1. Let students get the boxes they brought in so they can take them apart. They might also choose a box from the teacher's collection.
2. Let students begin taking the boxes apart. Some students will need help finding the seams. Help students identify the places where the boxes were glued together. If any boxes get torn, repair them with tape.
3. When the boxes are flat, have students trace the outline on newsprint or newspaper. Students may work in pairs, one student holding and one tracing.
4. Have students fold the boxes back into their original shapes. Provide tape to secure the flaps and seams.

   Ask questions while students work.

   - *How can you tell where the seams are?*
   - *What is the best procedure for tracing around the box?*
   - *What is the best way to put the box back together? Do you start with the top, the bottom, or the sides?*
   - *Compare the tracing and the reassembled box. How are they alike, and how are they different?*
   - *Can you find any flaps that were glued to hold the box together or tabs that can be tucked in a slot to reclose the box once it has been opened?*

MATH EXTENSION—MAKE A BOX

Investigation 2: Getting to Know Paper

- Cut around the shape, on solid line.
- Fold on dashed lines.
- Glue marked flaps.

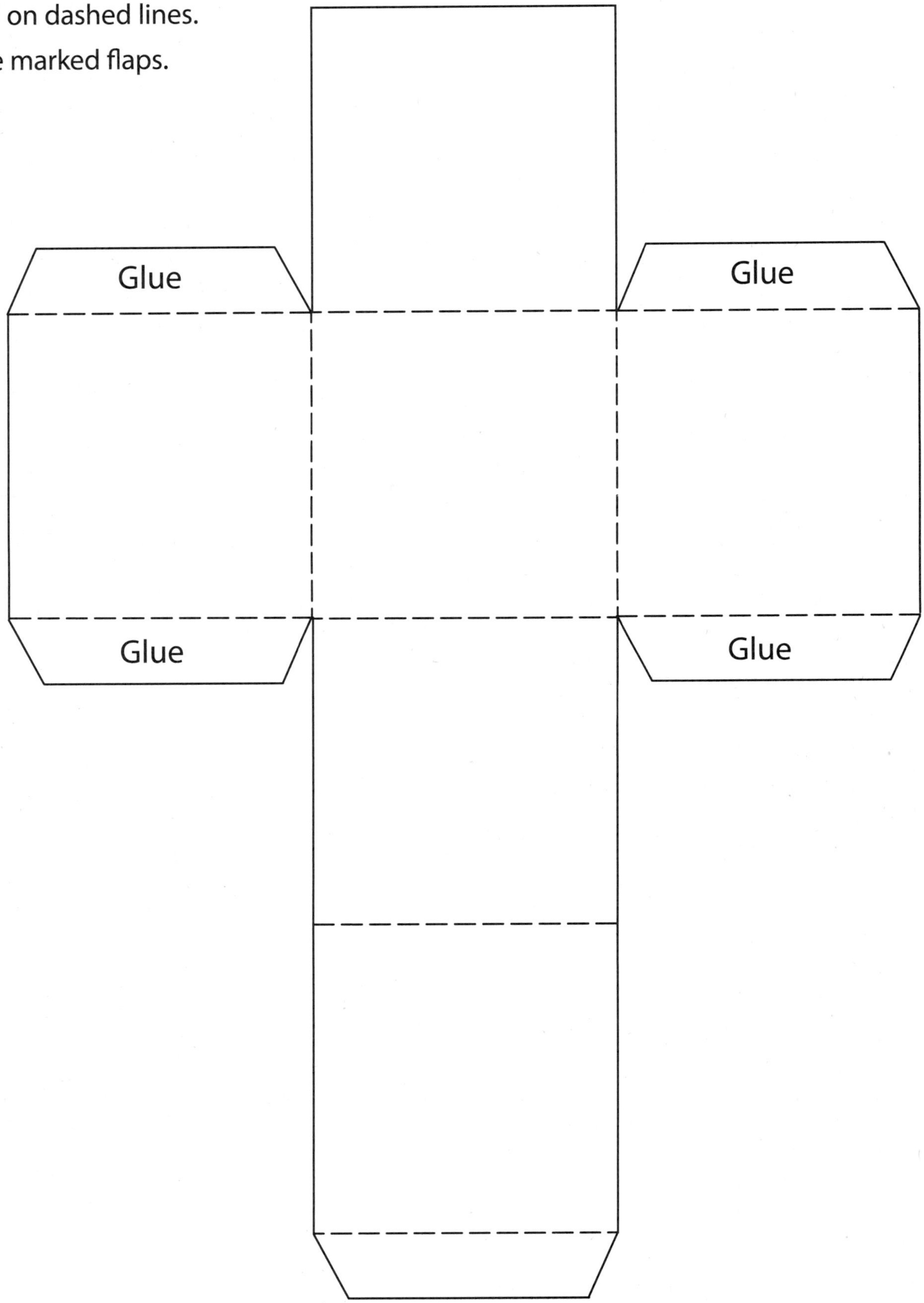

MATH EXTENSION—MAKE AN ENVELOPE

Investigation 2: Getting to Know Paper

- Cut around the shape on solid line.
- Fold on dashed lines.
- Glue marked flaps.

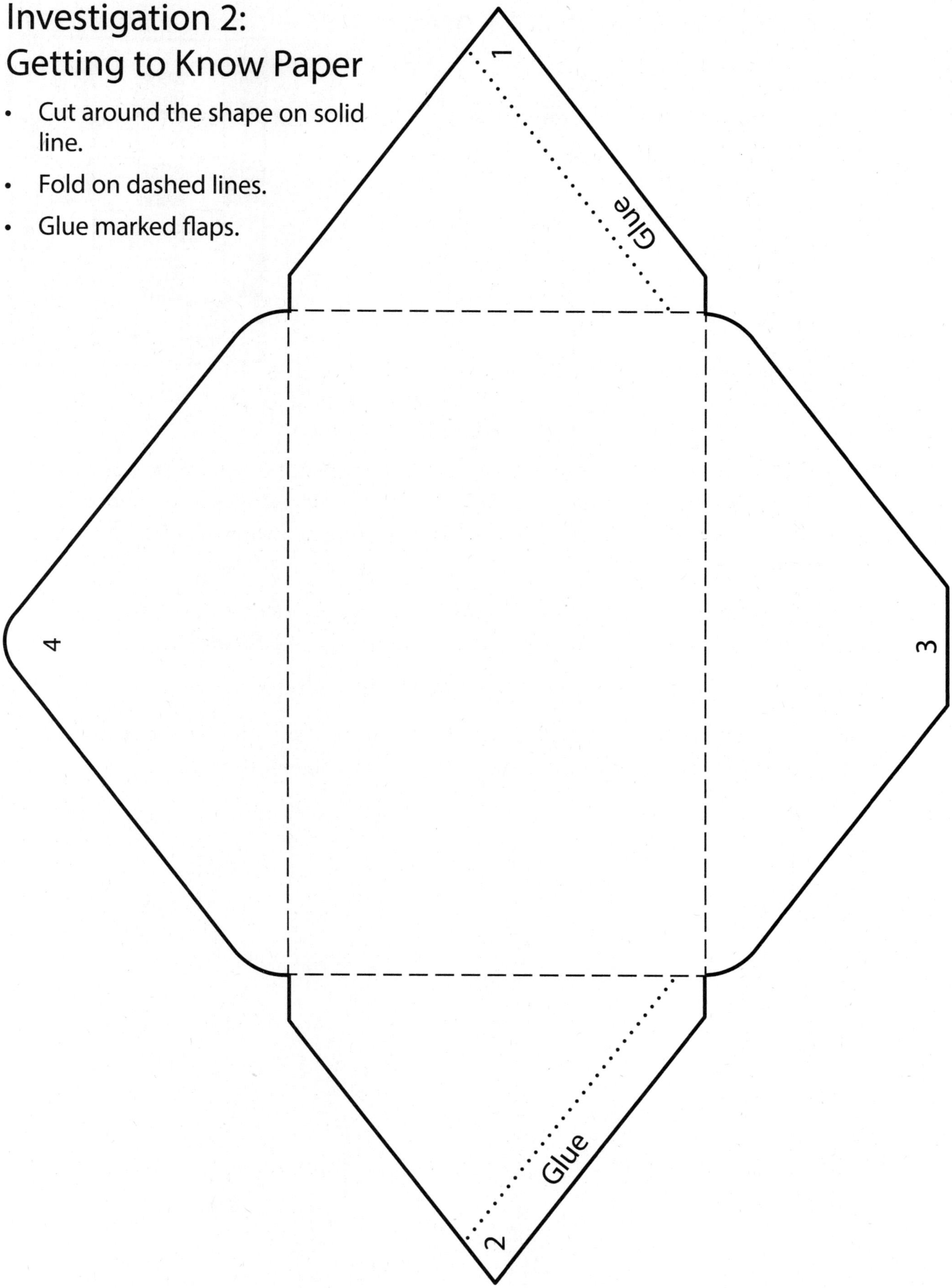

ART EXTENSION—PAPER WEAVING

Investigation 2: Getting to Know Paper

Materials

Construction-paper bases with five slits
Strips of construction paper, two colors
(different from base)
Glue
Scissors
Tape

Construction

NOTE: If students have not had any experience with weaving, they will need to work very slowly, step-by-step. If they have had experience, let them work independently.

1. Have students orient the base so the slits are going up and down. Weave the first colored strip under, over, under, and so on. Push the strip down to the bottom of the base. (The strips are longer than the base, so students won't accidentally pull them all the way through.)
2. Weave the other color of strip next. This time, students must start the strip by weaving over, then under. (Alternating the pattern is the difficult part.)
3. Continue weaving the strips until the entire paper is covered.
4. Students can put a drop of glue under both ends of each strip to attach it to the base. Students can then trim the woven strips to the size of the base.

ART EXTENSION—PIRATE HATS

Investigation 2: Getting to Know Paper

Materials

Sheets of newspaper
Stapler

Construction

1. Lay out a full sheet of newspaper. (Using two sheets makes the hat a bit sturdier.)

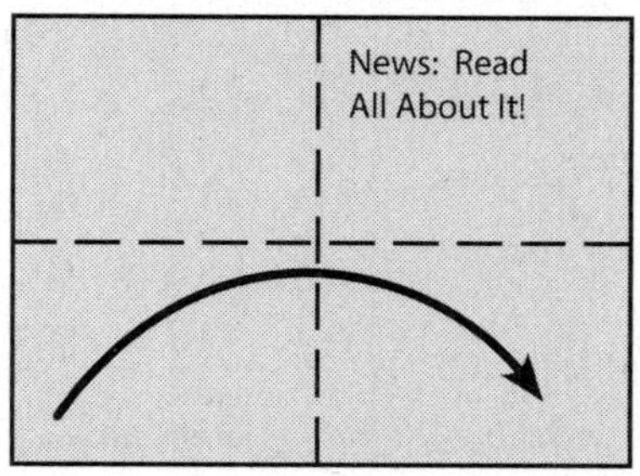

2. Fold the paper in half and rotate the folded paper 90 degrees.

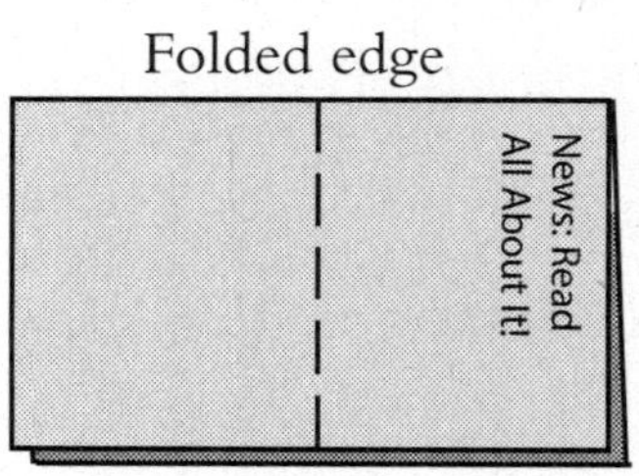

3. Fold down each of the upper corners, bringing the points to the center line.

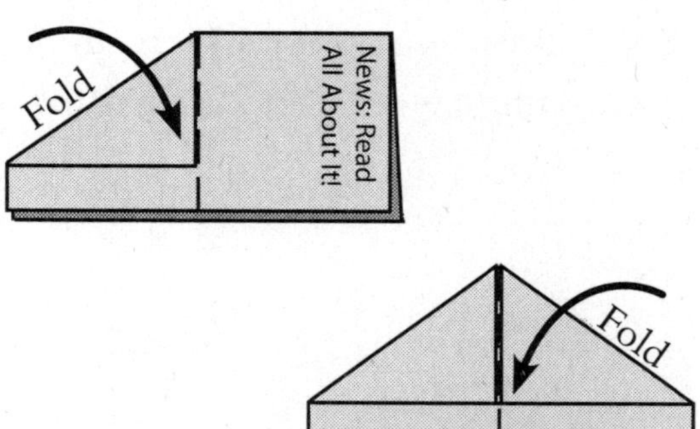

4. Fold up the bottom edge along the line created by the corner folds.

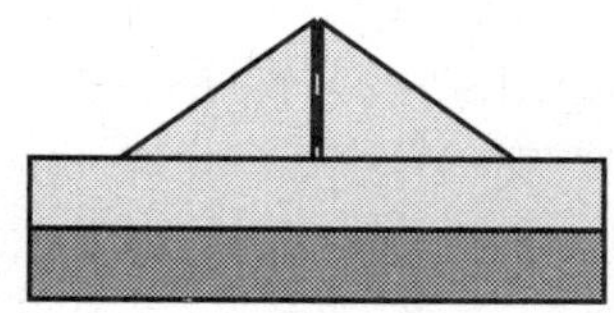

5. Flip the hat over, and fold up the other edge. Staple the flaps up in one or two places to keep the hat from unfolding and falling apart. But don't staple the hat closed!

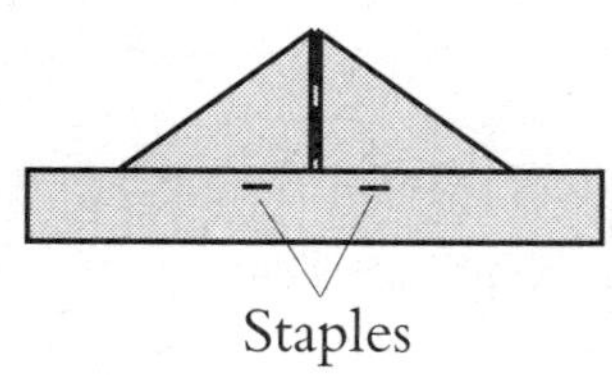

6. If the hat is too big, staple the hat together on each side, leaving enough room in the middle for the hat to fit comfortably.

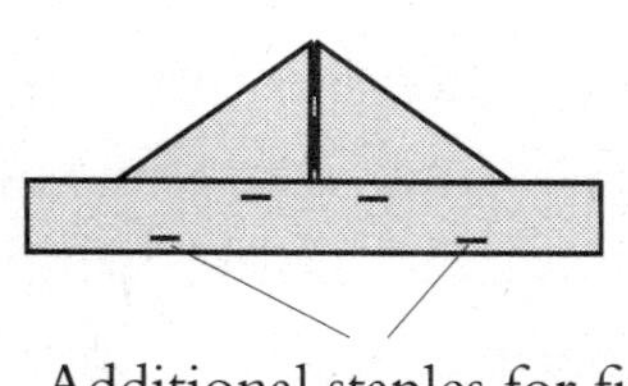

HOME/SCHOOL CONNECTION A

Investigation 2: Getting to Know Paper

At school, we have begun to investigate the properties of paper. Recently, we compared how easily different kinds of paper can be folded. Not all are alike! Here is a paper-folding project you can do together. If you have more than one kind of paper available, it would be interesting to make a cup from both papers and compare the two.

Materials

Drinking-cup pattern (Home/School Connection B)
Scissors

Construction

1. Cut the top off the pattern sheet.
2. Fold the sheet on the middle line that runs from corner to corner. The other lines should be on the outside, not folded in.

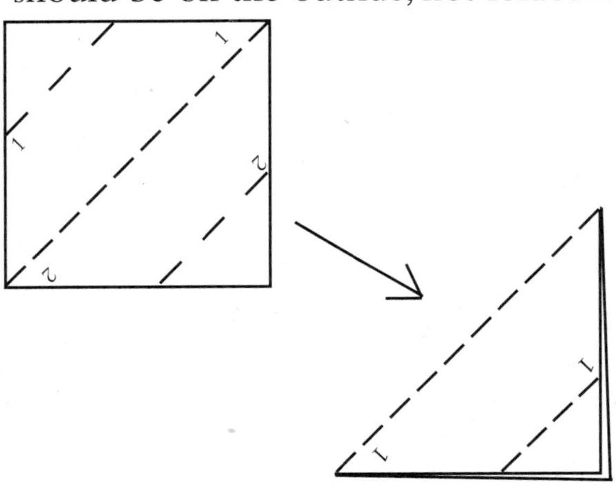

3. Lay the sheet on the table so the long side is closest to you, and the number 1s are showing.

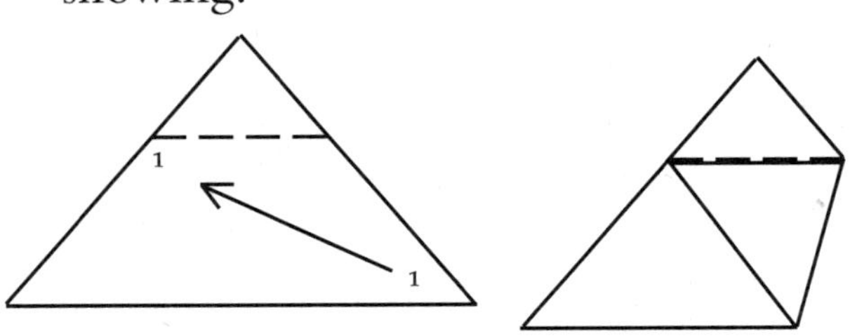

4. Fold the number 1 on the long side up to the number 1 in the middle of the other side. Fold so the edge of the paper is parallel to the short dotted line.
5. Flip the folded paper over and follow the same procedure, this time matching the 2s.

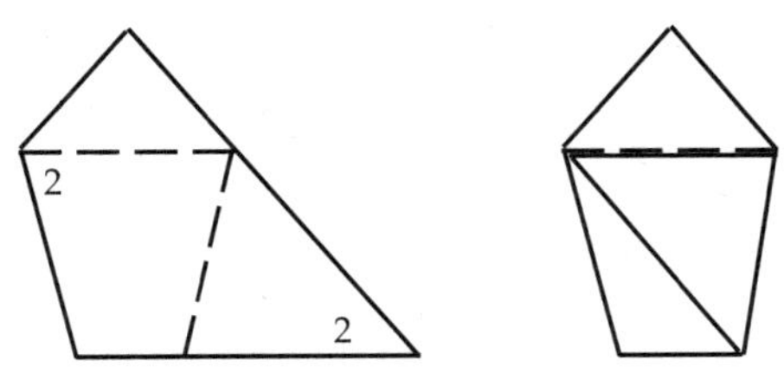

6. Separate the two triangle flaps at the top of the cup. Tuck one into each of the pockets formed by the previous two folds. Open the cup, fill it with water, and drink up!

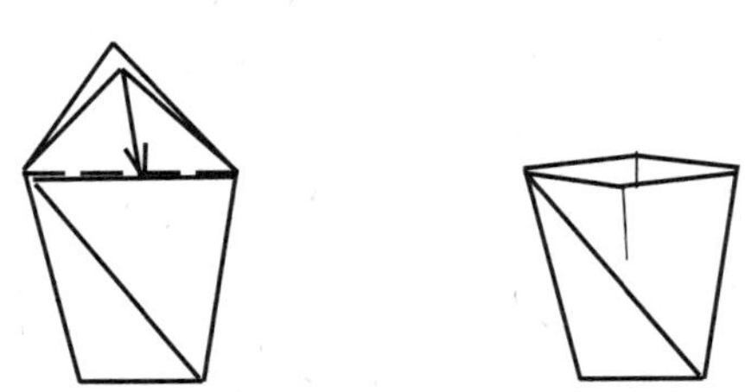

HOME/SCHOOL CONNECTION B

Investigation 2: Getting to Know Paper

Materials and Motion Module
Investigation 2: Getting to Know Paper
No. 27—Teacher Master

Cut on this line.

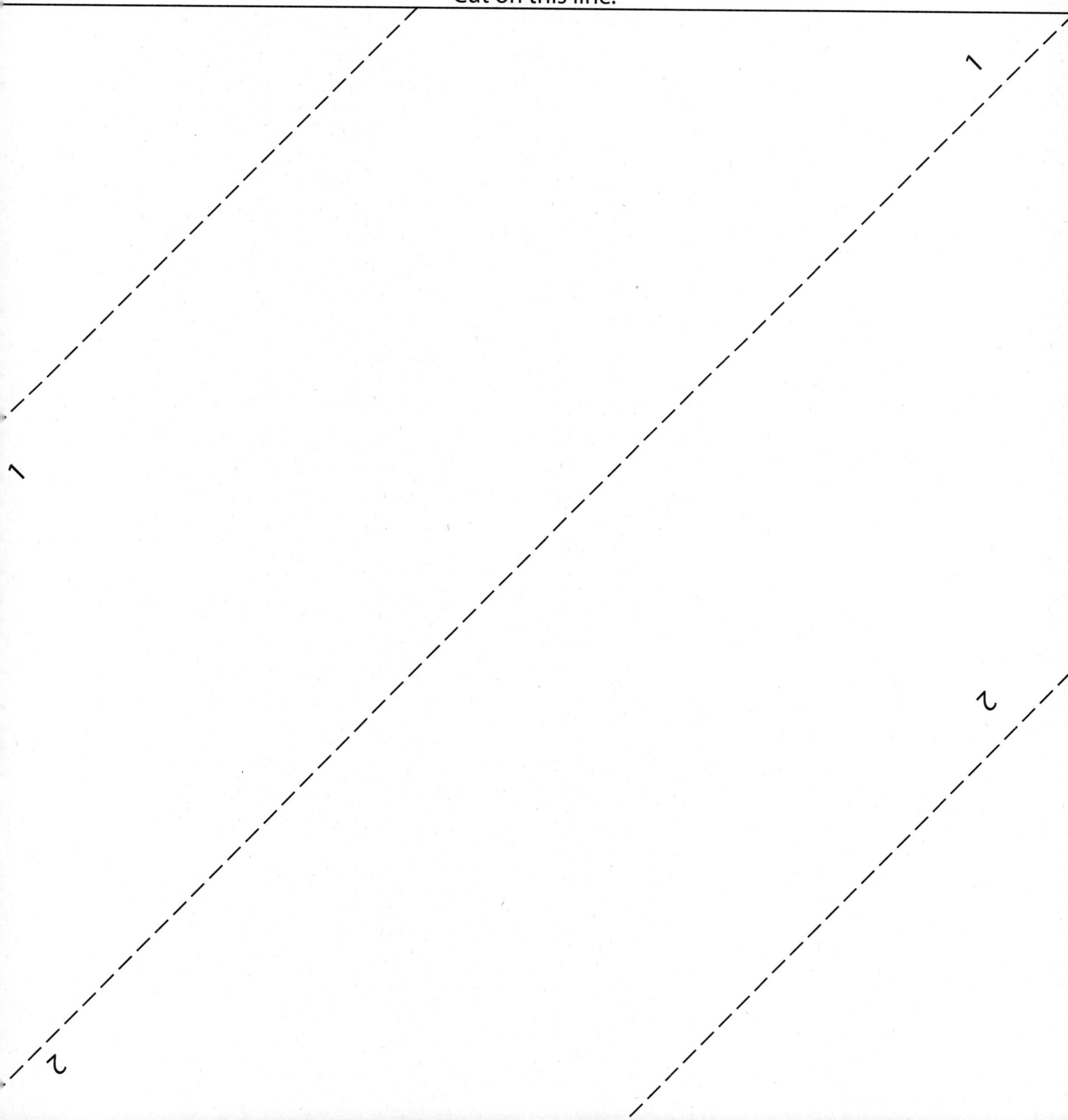

HOME/SCHOOL CONNECTION C

Investigation 2: Getting to Know Paper

Making a collage of different kinds of paper is a fun way to explore and compare the many properties of different papers. Your child will get the most from this activity if he or she has collected a wide variety of paper. Crepe paper, tissue paper, wallpaper, wrapping paper, and cardboard are all good choices. Little scraps are all that is needed, but take the time to gather an interesting collection.

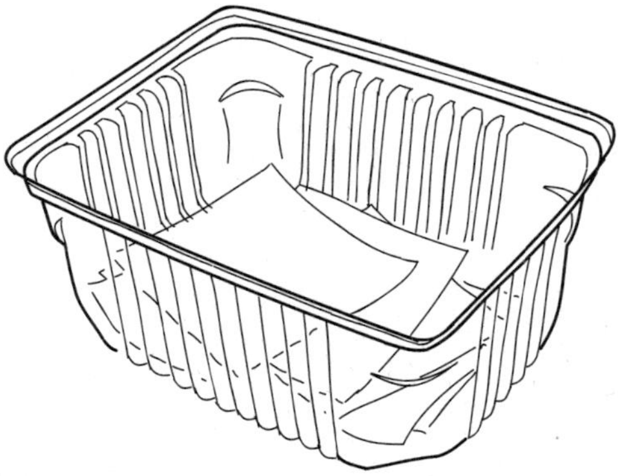

Materials

Variety of paper scraps
Glue
Scissors
Stapler (optional)
Construction paper

Construction

1. Cut a paper square to be used as a base for the collage. Experiment a bit together with different ways of using the paper scraps you have collected. Try tearing the paper; not all paper tears the same. Try curling different kinds of paper. There are many ways to bend or fold paper; accordion-folded paper makes nice pop-out effects. Sprinkle water on paper to see the effect; crepe paper has an especially interesting reaction to water drops.
2. Once the exploration has sparked ideas, let your child begin making a collage, and provide little guidance. Encourage him or her to cover the entire paper base. Here are some questions to guide discoveries.
   - *Are all the pieces of paper easy to cut or tear? Which are more difficult and why?*
   - *How many different kinds of paper are on your collage?*

Reminder

We will be making sculptures using paper and wood soon and could use any leftover scraps of interesting paper. We would appreciate any contributions for our collection. Thank you.

CENTER INSTRUCTIONS—FEELY BOXES AND FABRIC HUNT

Materials

Sets of five blue fabrics
Feely boxes

Set Up the Center

Put a set of five fabrics at each student's place. Make sure both students in a pair have the same set of five fabrics. Keep the feely boxes close by, ready to use.

Guide the Investigation

1. **Explore sets of fabric.** When students are settled, ask them to explore the pieces of fabric in front of them. Guide their exploration by asking questions.
   - *How are the fabrics different?*
   - *How do they feel?*
   - *Describe the texture of two different fabrics.*
   - *Can you match your fabrics to your partner's? How?*
   - *Do you think you could match the fabrics if you couldn't see them?*
2. **Add to the word wall.** As students offer their observations, add new or important vocabulary to the word wall.
3. **Demonstrate the activity.** Put a set of five different fabric samples in a feely box and a matching set of samples on the table. Sit in front of the feely box (the side with no hole). Show students how to put their hands through the holes in the sides to feel the fabrics inside the box.

   Ask one student to choose a fabric from the set outside the box and put it in your hands, inside the box. Find the fabric that matches by feeling the pieces in the box. When you find a match, take both pieces of fabric out of the box and check.

   If the fabrics don't match, try again. If they do match, give one fabric back to the student helper, and put the other one back in the feely box. Let students begin.
4. **Switch places.** After the "feeling" students have matched all five of the fabric pieces, have students switch places. You can also switch sets of fabric so everyone gets to work with all ten fabrics.
5. **Wrap up the session.** Ask students to explain how they were able to tell the fabrics apart. Ask which fabrics were easy to tell apart and which were difficult.

Vocabulary

Try to include these words in discussions with students:
fabric, nubby, rough, scratchy, shiny, slippery, smooth, soft, texture

FABRIC LABELS

This is made of **fabric.**	This is made of **fabric.**
This is made of **fabric.**	This is made of **fabric.**
This is made of **fabric.**	This is made of **fabric.**
This is made of **fabric.**	This is made of **fabric.**
This is made of **fabric.**	This is made of **fabric.**
This is made of **fabric.**	This is made of **fabric.**
This is made of **fabric.**	This is made of **fabric.**

CENTER INSTRUCTIONS—TAKING FABRIC APART

Materials

Burlap squares
Wool-plaid squares
Index cards
White glue
Loupes/magnifying lenses

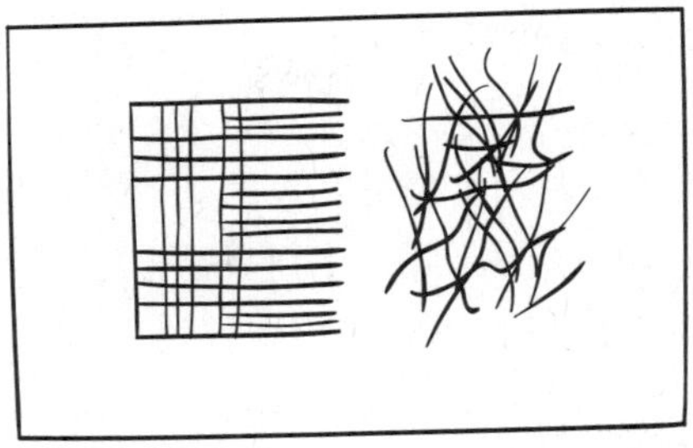

Set Up the Center

Place the fabric pieces, glue, index cards, and loupes/magnifying lenses in a convenient location for easy distribution.

Guide the Investigation

1. **Take apart burlap.** Give each student a small burlap square. Have students take the fabric apart, pulling off one thread at a time. Have them save the threads as they pull them off. Have them pull off only half the threads, so that a portion of the fabric sample is intact to glue onto the index cards.

2. **Ask questions to guide the discussion.**

   - *Can you see how the threads are woven together to make fabric? How do they come apart?*

   Show students the loupes/magnifying lenses. Let students get a close look at the threads. Ask,

   - *What do the threads look like when you see them through a magnifying lens?*

3. **Add to the word wall.** As students offer their observations, add any new or important vocabulary to the word wall.

4. **Glue threads and fabric to a card.** Have students glue onto an index card, the partial square of burlap and the unraveled threads they took off.

5. **Take apart the plaid.** Have students follow the same procedure with the wool plaid, using a second card to make another cloth-and-thread display. Discuss how the plaid is like the burlap and how it is different.

6. **Clean up.** Get the center ready for the next group.

Vocabulary

Try to include these words in discussions with students:
thread, warp, woof, woven

CENTER INSTRUCTIONS—WATER AND FABRIC

Materials

Sets of ten blue fabrics
Droppers
Water in cups
Paper towels
Clothesline and clothespins

Set Up the Center

Put a dropper at each student's place. Put a cup of water and a bag of blue fabric samples in a convenient location for each pair of students.

Guide the Investigation

1. **Drop water on terry cloth.** Have students lay a piece of terry cloth on the table. Ask what they think will happen if they put drops of water on the cloth. Have students squeeze a few drops of water onto the terry cloth. Discuss what they see.

2. **Compare nylon and satin.** Have students drop water on satin and nylon and observe what happens. Compare how the water acts on the different kinds of fabric.

3. **Add to the word wall.** As students offer their observations, add any new or important vocabulary to the word wall.

4. **Continue observing.** Encourage students to drop water on the other pieces of fabric in the set. Ask questions to guide discussion.
   - *Does the water soak into the fabric? Bead up on top? Bead up first and then soak in?*
   - *Which fabrics absorb the most water?*
   - *How does the water move through the fabric? Does it stay in one spot or spread out?*
   - *How does the fabric feel after you have one drop of water on it? Does it feel dry or wet?*

5. **Discuss fabric uses.** Discuss possible uses for fabrics that soak up water and those that are waterproof.
   - *What kind of fabric would be good for a raincoat? For a bath towel?*

6. **Hang the fabrics out to dry.** Hang the fabrics on the clothesline to dry. Use clothespins to hold them securely on the line.

7. **Clean up.** Get the center ready for the next group.

Vocabulary

Try to include these words in discussions with students:
absorb, bead up, dropper, least, most, soak, spread, waterproof

CLOTHING PICTURES

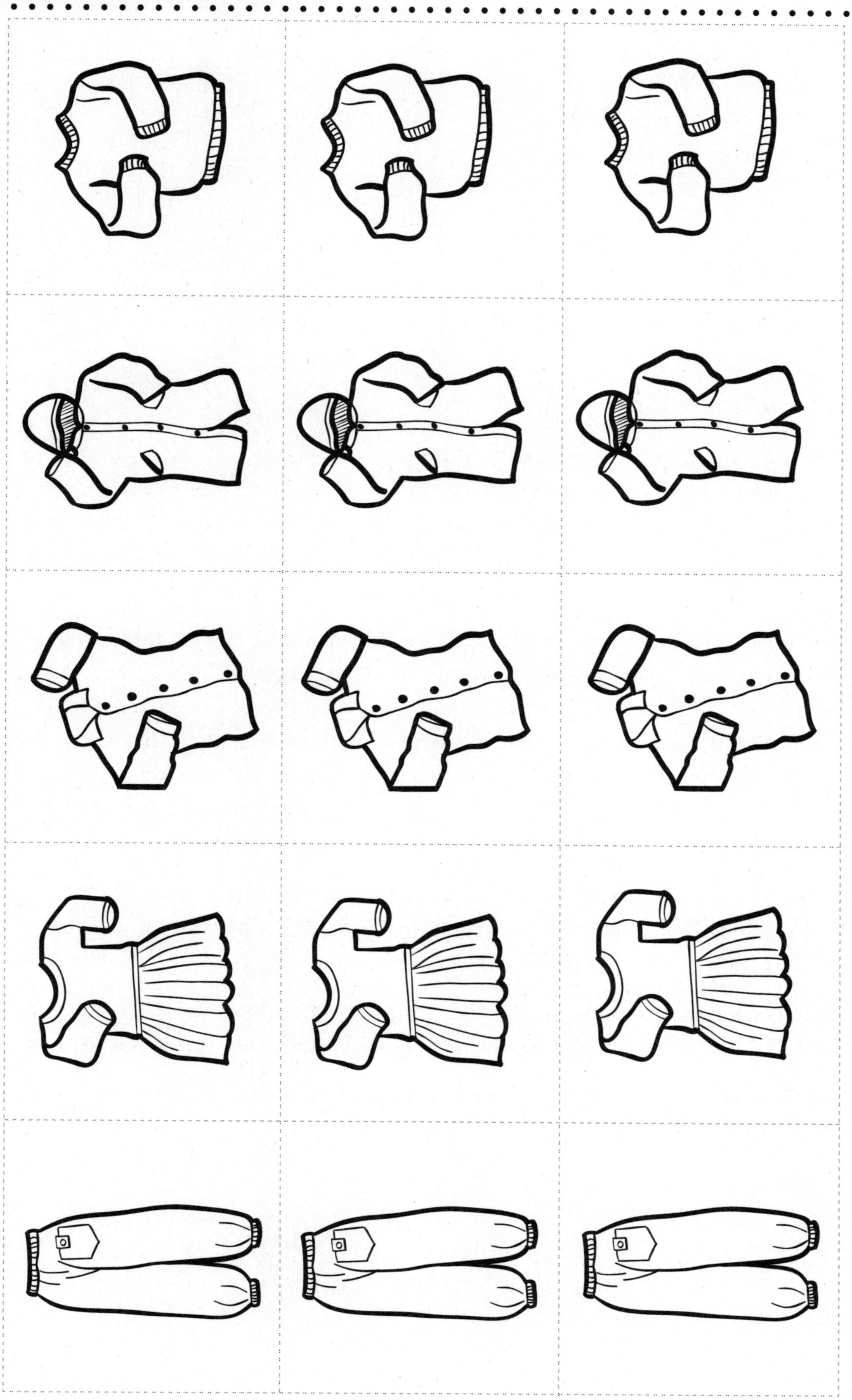

CENTER INSTRUCTIONS—BUILDING STRUCTURES

Materials

Paper scraps
Fabric scraps
Newspaper mats
White glue
Paper plates
Wood scraps
Pencil
Craft sticks
Scissors

Set Up the Center

Have all materials available at the center for students to use as needed.

Guide the Investigation

1. **Review the challenge.** Remind students that their job is to construct something out of wood, paper, and fabric. It should be designed to keep a cup of water cool in the sunshine. Point out the many kinds of paper, scrap wood, fabric, and craft sticks to use to make a structure.
2. **Begin construction.** Let students work freely for a time before providing glue. Tell them to pile and hold pieces of wood and paper together in various ways to figure out how they would like to assemble their structures. Let the construction begin. If students have difficulty getting started, make some suggestions that include two or more materials. Remind students to leave a space in their structures to place the cup of water. Encourage students to keep their structures on a paper plate if possible.
3. **Label the structures.** While students work, write each student's name and the date on a craft stick. Students should include their craft sticks as part of their structures so they can be identified later.
4. **Dry the structures overnight.** Set the structures in a place where they can dry overnight. Check to make sure each one is labeled with the student's name.
5. **Add to the word wall.** Add new or important words that students use to the word wall.
6. **Clean up.** Get the center ready for the next group.

Vocabulary

Try to include these words in discussions with students:
cloth, construction, design, fabric, structure

Investigation 3: Getting to Know Fabric

At school, we have been using our senses to observe the many ways that fabrics differ. We have also looked around to see where fabrics are used. Here are two ways to practice the vocabulary and skills your child is developing.

- Give a clue: "I'm thinking of something that is made of fabric, and it is [blue, soft, striped]." Take turns describing and identifying fabrics in use around your home.
- Have your child search for four different ways that fabric is used around your home or neighborhood. Have him or her draw and label the four uses for fabric.

Investigation 4: Getting Things to Move

In science class we have been studying things that move. We have investigated things that roll, like balls, wheeled toys, and other round objects. We are also interested in other kinds of motion—swinging (pendulums and play swings) and spinning (pinwheels and fans).

Take a few minutes to tour your home with your student to make a list of objects and systems that display one or more classes of motion. Talk about how things move.

Another activity you could engage in together would be to look through some catalogs and newspaper advertising circulars for images of moving objects—home appliances (mixers, blenders, clothes washers and driers), tools (drills, saws, and sanders), and toys (skateboards, wagons, bicycles, etc.). Cut out the images and make a "things that move" collage to take to school to share.

Assessment Masters

Assessment Checklist A Disciplinary Core Ideas

Materials and Motion, Investigation ____________

Start date ____________

End date ____________

Student names	**PS1.A: Structure and properties of matter** Matter can be described and classified by its observable properties.	**PS3.B: Conservation of energy and energy transfer** Sunlight warms Earth's surface.	**ESS3.A: Natural resources** Living things need water, air, and resources from the land; they live in places that have the things they need. Humans use natural resources for everything they do.	**ESS3.C: Human impacts on Earth systems** Things that people do to live comfortably can affect the world around them. But they can make choices that reduce their impacts on the land, water, air, and other living things.

Assessment Checklist B Disciplinary Core Ideas

Materials and Motion, Investigation ________

Start date ________________

End date ________________

Student names	**PS2.A: Forces and motion** Pushes and pulls can have different strengths and directions. Pushing or pulling on an object can change the speed or direction of its motion and can start or stop it.	**PS2.B: Types of interactions** When objects touch or collide, they push on one another and can change motion.	**PS3.C: Relationship between energy and forces** A bigger push or pull makes things go faster.

Assessment Checklist C Disciplinary Core Ideas

Materials and Motion, Investigation ________

Start date ________

End date ________

Student names	**ETS1.A: Defining and delimiting engineering problems** A situation that people want to change or create can be approached as a problem to be solved through engineering.	**ETS1.B: Developing possible solutions** Designs can be conveyed through sketches, drawings, or physical models.	**ETS1.C: Optimizing the design solution** Because there is always more than one possible solution to a problem, it is useful to compare and test designs.

Assessment Checklist D Science and Engineering Practices

Materials and Motion, Investigation ________

Start date ____________

End date ____________

Student names	Asking questions and defining problems	Developing and using models	Planning and carrying out investigations	Analyzing and interpreting data	Using mathematics and computational thinking	Constructing explanations and designing solutions	Engaging in argument from evidence	Obtaining, evaluating and communicating information

Assessment Checklist E Crosscutting Concepts

Materials and Motion, Investigation ________

Start date ____________

End date ____________

Student names	Patterns	Cause and effect	Scale, proportion, and quantity	Systems and system models	Energy and matter	Structure and function

NARRATIVE REPORT—MATERIALS AND MOTION

Student's Name ______________________________ Date from __________ to _________

PURPOSE OF THE MODULE

FOSS Materials and Motion Module provides young students with opportunities to investigate motion (the push or pull that starts and stops moving objects, and changes speed and direction) and to observe, compare, and transform the physical properties of materials made from natural resources—objects made from wood, paper, and fabric. Students look at many properties of these materials (their size, color, texture, flexibility, buoyancy, and absorbency), how products are put together, how materials change when mixed with or put in water, and how materials can be changed by human action or engineering (sanding, tearing, cutting, and gluing). Students take on the role of engineers as they design systems for rolling a ball to achieve a specific outcome; making particleboard and plywood; and making new paper from old paper. As an engineering design project, students select materials and design a shelter to reduce the warming effect of sunlight on Earth. Throughout the module, they use science and engineering practices by making and communicating observations, making comparisons, and looking for patterns and cause and effect. Students set up a recycling center and sort objects based on what materials they are made of and, in so doing, learn about conserving Earth's resources.

Comments: ____

____

____

____

____

____

____

____

____

____

____

____

____

NGSS CHECKLIST A Materials and Motion Module

	Student progress
Disciplinary core ideas	
PS1.A: Structure and properties of matter Matter can be described and classified by its observable properties.	
PS2.A: Force and motion Pushes and pulls can have different strengths and directions. Pushing or pulling on an object can change the speed or direction of its motion and can start or stop it.	
PS2.B: Types of interactions When objects touch or collide, they push on one another and can change motion.	
PS3.B: Conservation of energy and energy transfer Sunlight warms Earth's surface.	
PS3.C: Relationship between energy and forces A bigger push or pull makes things go faster.	
ESS3.A: Natural resources Living things need water, air, and resources from the land, and they live in places that have the things they need. Humans use natural resources for everything they do.	
ESS3.C: Human impacts on Earth systems Things that people do to live comfortably can affect the world around them. But they can make choices that reduce their impacts on the land, water, air, and other living things.	
ETS1.A: Defining and delimiting engineering problems A situation that people want to change or create can be approached as a problem to be solved through engineering.	
ETS1.B: Developing possible solutions Designs can be conveyed through sketches, drawings, or physical models.	
ETS1.C: Optimizing the design solution Because there is always more than one possible solution to a problem, it is useful to compare and test designs.	

NGSS CHECKLIST B Materials and Motion Module

	Student progress
Science and engineering practices	
• Asking questions and defining problems	
• Developing and using models	
• Planning and carrying out investigations	
• Analyzing and interpreting data	
• Using mathematics and computational thinking	
• Constructing explanations and designing solutions	
• Engaging in argument from evidence	
• Obtaining, evaluating, and communicating information	
Crosscutting concepts	
• Patterns	
• Cause and effect	
• Scale, proportion, and quantity	
• Systems and system models	
• Energy and matter	
• Structure and function	